Plastic Bottles
Processing, Recycling, Regulations and Alternatives

Tanvi Vats and Shubhanshi Sharma

CRC Press
Taylor & Francis Group
Boca Raton London New York

CRC Press is an imprint of the
Taylor & Francis Group, an **informa** business

First edition published 2025
by CRC Press
2385 NW Executive Center Drive, Suite 320, Boca Raton FL 33431

and by CRC Press
4 Park Square, Milton Park, Abingdon, Oxon, OX14 4RN

CRC Press is an imprint of Taylor & Francis Group, LLC

ISBN: 978-1-032-89084-5 (hbk)
ISBN: 978-1-032-89086-9 (pbk)
ISBN: 978-1-003-54110-3 (ebk)

DOI: 10.1201/9781003541103

Typeset in Times
by codeMantra

Dedicated to the Almighty God for all his bliss…

Contents

Plastic Bottles

Plastic Bottles: Processing, Recycling, Regulations and Alternatives explores the lifecycle of plastic bottles, from creation to disposal, offering a comprehensive and accessible look at bottle packaging. This book is divided into five parts as follows:

- **Part 1: The History of Plastic Bottles**
  - Traces the development and integration of plastic bottles into daily life.
  - Covers materials used, labeling, and manufacturing processes in the industry.
- **Part 2: Environmental Impact**
  - Examines the limitations of plastic bottles and their environmental consequences.
  - Discusses challenges in recycling and showcases case studies.
  - Highlights advanced recycling technologies and techniques.
- **Part 3: Biopolymers as an Alternative**
  - Introduces biopolymers as sustainable alternatives to traditional plastic.
  - Explores types of biopolymers suitable for bottle production.
  - Discusses potential benefits and challenges of biopolymer adoption.
- **Part 4: Regulations and Policies**
  - Focuses on global regulatory frameworks for plastic and biopolymer use.
  - Covers Extended Producer Responsibility (EPR) and its role in waste management.
- **Part 5: Future of Bottle Packaging**
  - Looks ahead at developments in sustainable packaging solutions.
  - Discusses innovations in recycling and emerging research trends.

Target Audience

✓ Accessible to students, academics, and industry professionals from both scientific and non-scientific backgrounds.
✓ Concise and easy-to-read, making it suitable for a wide ranging audience.

This book provides a thorough yet compact overview of the plastic and bio-polymer bottle packaging industry, offering valuable insights for both academia and industry.

Author Biography

Dr. Tanvi Vats completed her PhD in chemistry jointly from GGS Indraprastha University and National Physical University, New Delhi. She currently serves Gautam Buddha University, Greater Noida as a faculty member where she teaches BTech, BSc, and MSc students. Her research interests include nanomaterials and science-related aspects of Intellectual Property Rights. She has attended various national and international conferences and has several research papers, book chapters, encyclopedia articles, and popular science articles to her credit. She has co-authored two books, *Microwave assisted Polymerization* (published with RSC, Cambridge, UK) and *Comprehensive Chemistry: Theory and Practice* (published with Viva Books Private Limited, New Delhi). She is a life-time member of the Association of Environmental Analytical Chemistry of India (AEACI), Analytical Chemistry Division Bhabha Atomic Research Centre, Trombay, Mumbai, and the "Academy of Microscope Science & Technology," DMSRDE, Kanpur, India.

Shubhanshi Sharma is an aspiring researcher in the field of polymer packaging. She has completed her MSc in Applied Chemistry from Gautam Buddha University, Greater Noida, India. It was during her Master's thesis on Bioengineering and biopolymers used for tissue engineering and regenerative medicine that she developed an interest in polymer packaging and also an aptitude in the area of science communication. She has a popular science article entitled "Packaging Food: Looking for Durable & Efficient Options" published in *Science Reporter*.

During her leisure hours she finds great enjoyment in delving into books and documentaries that explore the rich tapestry of Indian history. Additionally, she finds solace in mountains and frequently goes trekking with friends and family.

Plastic Bottles
A Necessary Evil?

1

CHAPTER 1.1: INTRODUCTION: THE OMNIPRESENT PLASTIC BOTTLE

It is difficult to imagine the contemporary world devoid of any plastic bottles. They have so seamlessly merged into our lives that most of the time their presence goes unnoticed. Starting from the morning tetra pack of the milk we encounter various bottles in our everyday life. Some of them include packaging for shampoos, household cleansers, beverages, oils, so on and so forth. The emergence of plastic bottles marked the beginning of an era of a benign, lightweight, long-lasting, and a multi-faceted solution for storing and transporting liquids. However, the lack of human sense of balance and limit has led to their rampant overuse, which, in turn, has made them a major nuisance. This misuse and overuse of plastic bottles has made them a great concern for the sustainable development worldwide.

1.1.1 The Rise of Plastic Bottles

The need for a material better than glass or metal in terms of versatility and cost-effectiveness led to the development of synthetic polymers in the early 20th century. The 1940s saw the invention of polyethylene terephthalate (PET), which paved the way for creation of first plastic bottles. This was initiation of a sort of revolution, over the time with the advancements in manufacturing techniques the versatility of the bottles increased and so did the consumer demands. The ease of use and availability resulted in the rapid propagation of plastic bottles across every nook and corner of the world.

Initial forms of plastics were mainly celluloid and Bakelite and they were synthesized using organic materials like cellulose and phenols. These semi-synthetic materials paved the way for fully synthetic plastics. Several types of plastics with different properties were synthesized during this period, however, it was the synthesis of PET, (Polyethylene Terephthalate) that was a milestone in the bottle industry. PET had various properties which worked for it, PET was lightweight, transparent and break resistant. By the 1970s PET bottles were fully projected as a light weight, break resistant alternative to glass bottles. This revolutionized packaging industry and one of the components of its popularity has been the reduced logistic costs as compared to the heavier and fragile glass bottles.

1.1.2 Convenience Meets Consumption

The major beneficiary of the so-called plastic bottle revolution is, unanimously the beverage industry. The convenience offered by the plastic bottles is unparalleled. The ease of transportation, the break free material, the bright colours they can come into and the plethora of sizes they can be made available has indeed given the consumers a customized experience with the plastic bottles. The beverage industry milked this opportunity and started the mass production and distribution of bottled drinks. Other industries also followed the suit and in the matter of time the plastic bottles became an integral part of modern life.

An interesting observation in this case is the integration of bottled water as an essential item of our day-to-day life. In the late 20th century, the bottled water was a lifestyle product which was pure and healthy affordable only by a certain class of society. The advent of plastic technology cheap and affordable water packaging made it an essential everyday product. A similar shift was observed in the soft drinks industry. The portability of plastic bottles ensured that they reach all the possible market avenues catering to the needs of people across all the strata of society.

1.1.3 Environmental Impact

The introduction of plastic bottles has definitely come with multiple advantages, yet the environmental impact caused by their overuse cannot go unheeded. The production of bottles has a heavy dependence on fossil fuels leading to greenhouse gas emissions. The post use disposal of the plastic bottles is another challenge. Though a fraction of bottles used are recycled, a huge percentage of them end up in landfills, incinerators, and also as a litter in natural environment. It is no surprise that even the most pristine locations like remote villages on the highest of the mountains or deepest of the sea or in the densest of the forests have a plastic bottle lurking from somewhere. The same durability which makes plastic bottles so desirable makes them persist in the environment for indefinite period of time causing immense harm to nature and wildlife.

The lifeline of plastic bottles involves a trail of ecological footprints. The synthesis of plastic resins starting from the extraction and processing of raw materials results in the significant carbon dioxide emission. Post-production, as these bottles have the economical logistics, they travel very long distances hence have additional carbon foot prints. The used bottles should ideally be recycled but due to inadequate waste management a humungous fraction of them end up in landfills or in the oceans. Ultimately the consequences on the human race have also started showing up. There are now studies showing the infiltration of microplastics in human food chains raising concerns about long-term health impacts.

1.1.4 Plastic Bottles and Sustainable Development

As plastic bottles have become so ingrained in our lives the challenges posed by them require a multi-layered solution. This book investigates innovative solutions that mitigate the environmental impact of plastic bottles. The work discusses various strategies aimed at creating a more sustainable future. These include advanced recycling techniques, development of environmentally benign materials and policy initiatives among various economies.

One of the key counters offered by the plastic bottle advocates is "bottle recycling." However, recycling is not devoid of the challenges. The major challenges are contamination, inadequate sorting and economic challenges in recycling. In the proceeding chapters we will be discussing the advanced recycling methods which promising address the recycling challenges. We will also discuss more sustainable options for the plastic bottles which are not only ecologically safe but also economically well placed.

Another area we can work on is consumer awareness programmes. Significant changes in the environment can be attained by altering consumer behaviour. Educating the consumers on the environmental impact of the bottles and adoption of more ecofriendly alternatives can actually make a huge difference.

1.1.5 Join the Journey

In this book we tag along the path of the plastic bottles right from their inception to their becoming an integral, omnipresent part of our lives. The book ambles along the past, present, and future of one of the most noteworthy inventions of modern era.

The work touches upon all the aspects of plastic bottles starting from their composition, labelling, manufacturing process, their effect on the planet, regulatory framework, policy matters, and the future possibilities. The potential application of biopolymers in the bottle industry has also been covered.

The aim of this book is to create an awareness and deeper understanding of plastic bottles and their environmental impact so that consumers can make an informed decision. The choice between sustainability and development should not be exclusive of each other rather they should complement each other. We hope that this book acts as a catalyst for change and encourage all of us to rethink our relationship with our planet and together we strive for sustainable future.

CHAPTER 1.2: COMPOSITION: MATERIALS USED

Historically, polymeric polymers and glassware have been the primary materials used in the fabrication of water and beverage bottles. These materials combine high electrical, mechanical, and thermal qualities with exceptional chemical resistance and a stable dimensional structure [1]. Because they are inexpensive, have robust tensile characteristics, and function as a powerful defence against oxygen, carbon dioxide, and water vapour, petrochemical plastics are primarily employed in the bottle packaging industry. These polymers come in a large range and are utilized in packaging in both flexible and rigid forms. These polymers fall into one of two categories: thermosets or thermoplastics. Heat can be used to process and reprocess thermoplastics. And since they can be reprocessed, this class of polymers can be recycled because they are easily moulded into various shapes, which makes them better suited for food packaging. The thermoplastics most often used in bottle packaging materials are low-density polyethylene (LDPE), polypropylene (PP), polyvinyl chloride (PVC), polyethylene terephthalate (PET), high-density polyethylene (HDPE), polystyrene (PS), and expanded polystyrene [2]. Some of the commonly used materials used in the packaging of bottles are discussed below.

1.2.1 The Polymers in the Polyethylene Family

Polyethylene (PE) is a crucial raw material for the petrochemical sector. Among the five major synthetic resins, it is also the species with the highest import volume and production capacity among domestic synthetic resins [3]. Polyethylene, also expressed as $[CH_2–CH_2]_n$, is a long-chain ethylene ($H_2C=CH_2$) polymer that is chemically generated through ethane polymerization. Its side chains, which can be added based on the manufacturing method, give it a significant deal of variation [4].

For convenience, polyethylenes are generally categorized according to their density. The more the branching in the structure, the lesser these chains can be packed together (which means a greater steric hindrance) and therefore the lower the density of the material [5]. Various types of polyethylene have been identified based on the age of industrialization: low-density polyethylene (LDPE) with a density of 0.91~0.925 g/cm³ through high-pressure polymerization at 150~200 MPa; low-pressure polymerization of HDPE is the second generation of polyethylene; in 1977, the performance of linear low-density polyethylene (LLDPE) was produced industrially with many characteristics [3].

1.2.1.1 High-Density Polyethylene (HDPE)

HDPE is a polyethylene with a density of about 0.946–0.976 g/cm³ [3]. It consists of a smaller number of branched chains and is prepared through a low-pressure method and Ziegler-Natta or Phillips initiators to control the chain formation, resulting in a highly unbranched structure. Because of its larger crystalline content compared to LDPE, it is more stiff, opaque, and has a higher tensile strength. The majority of HDPE's uses are in heavy-duty products like pallets, drums, crates, and intermediate bulk containers, as well as in rigid packaging like milk and household chemical bottles. It's also utilized for screw caps with a short lifespan, like those on milk and beverage bottles [5].

HDPE is highly resistant to a variety of chemicals because of which it can be used to store corrosive materials without running the danger of deterioration or contamination. Furthermore, because HDPE is sturdy, the bottles can endure the rigors of handling, storing, and shipping, offering a safe containment for a variety of chemical goods. Given that HDPE is so adaptable, it can be made into a wide range of sizes and forms to meet the unique requirements of different sectors. HDPE can be used for a wide range of purposes, from big agricultural containers to tiny medicinal vials [6].

1.2.1.2 Low-Density Polyethylene (LDPE)

Low-density polyethylene is the lightest of the polyethylene resins which is produced by a high-pressure method. Compared with high-density polyethylene

[3, 5, 6], it has a mix of long and short branched chains, lower crystallinity (around 50%–65%) and has a mix of long and short branched chains, making it translucent in appearance [7]. It is soft and flexible with good elongation before breakage and good puncture resistance [5], thereby making it an excellent candidate in the manufacturing of dispensing bottles, and squeezable packaging for several products such as shampoos, conditioners, and adhesives. It has good low-temperature resistance to −70°C as well as excellent chemical stability. It is resistant to acid, alkali, and salt aqueous solutions.

1.2.1.3 Linear Low-Density Polyethylene (LLDPE)

LLDPE is a tasteless, odourless, and non-toxic material having a density of 0.915–0.935 g/cm^3. While linear low-density polyethylene (LLDPE) lacks long-chain branching and has a shorter, non-uniform side chain distribution, it is comparable in density to linear low-density polyethylene (LDPE). Copolymerizing ethylene at low pressure with butene, hexane, octane, or 4-methyl pentene yields LLDPE [5, 8]. Comparing LDPE to LLDPE, LDPE has lesser transparency, a greater melting point, and better mechanical characteristics. Yet, because of the LLDPE's residual shortcomings in terms of strength, stiffness, and heat resistance, studies on this material have drawn a lot of attention. For this reason, adding nanofillers like clay, silica, and titania nanoparticles can further improve the thermal and mechanical characteristics of LLDPE [8]. Plastic bottles shrink wrap is frequently composed of LLDPE. Due to its resilience, LLDPE is also used to make the plastic can rings that keep multi-pack canned beverages together [9].

1.2.1.4 Ethylene Vinyl Alcohol (EVOH)

Composed of vinyl alcohol and ethylene monomer units, EVOH is a random copolymer with a semi-crystalline structure [5, 10]. It is one of the polymers that is frequently used in packaging that has the lowest oxygen penetration (<2 cc/m^2/day) that has been recorded [5, 11], however, because of the −OH groups, it becomes hydrophilic—that is, it draws water, lowering the oxygen barrier. EVOH needs to be "sandwiched" to keep moisture out in order for the oxygen barrier to function as intended. Co-extrusion is a popular method used to do this; PET/EVOH/PET bottles for sauce and mayonnaise are one example [5].

1.2.2 Polypropylene (PP)

The structure of polypropylene (PP) is similar to polyethylene but it has a methyl group on the carbon main chain [8]. Developed in 1954, polypropylene gained popularity quickly due to its lowest density among all commercial polymers. PP can be transformed using a number of methods, such as injection moulding and extrusion, and has a high chemical resistance. It is a catalysed polymer composed of propylene. The main benefit of PP is its high temperature resistance, which makes it ideal for products like trays, funnels, buckets, bottles, carboys, and instrument jars that need to be cleaned or sterilized regularly for use in a medical setting [12]. The most popular packaging materials for yoghurt tubs, bottle caps, coffee cups, and soft drink, water, or syrup bottles are made of PP [13].

Typically, opaque and low in density, polypropylene resin has superior thermoforming and injection moulding properties. In the bottle market, it mainly rivals polyethylene since it can be produced transparent for see-through applications, whereas polyethylene can only be made translucent, such in milk jugs. While polypropylene performs rather well, it cannot equal the visual clarity of polymers like polycarbonate. It works well for extrusion and moulding applications, including blow moulding, because of its low viscosity at melt temperatures. Pill bottles and similar items are made of polypropylene (PP) [14].

1.2.3 Polycarbonates (PC)

Bisphenol A ($C_{15}H_{16}O_2$) and phosgene ($COCl_2$) are polymerized to create polycarbonates [14]. It is a synthetic thermoplasticwhich is a choice for many uses in the society. It is a member of the amorphous thermoplastic class. Its mechanical, thermal, optical, and electrical properties are all distinct and mouldable. Applications in the fields of automotive, consumer products, medical devices, electrical and electronics, packaging, optical media, and civil engineering and construction as well as home and industrial use are all very beneficial [15]. Many mineral acids, alcohols, mild soaps, petroleum oils (although not all oil additives), silicone oils and greases, and low alkali concentrations have been shown to be compatible with polycarbonate [16].

When considering other bottle-making polymers, PC is more expensive, hence its application is limited to expensive reusable bottles, like those used for nursing or found on water coolers or in laboratory environments. It is ideal for bottles that need to exhibit their contents with the clarity of glass but also be able to withstand frequent, occasionally rough handling because of its exceptional optical qualities and robustness. Compared to a glass bottle, a polycarbonate bottle is more robust and less prone to breaking. Polycarbonate bottles are a popular choice for reusable, lightweight, and unbreakable portable water bottles. Polycarbonate bottles, also referred to as sport bottles, are becoming more and more popular among customers who care about cutting waste. Polycarbonate containers provide a long-lasting alternative to single-use bottles of plastic that can damage the environment. It's an autoclavable material that can resist multiple washings [17].

1.2.4 Polyethyleneterephthalate (PET)

A thermoplastic resin of the polyester class, polyethylene terephthalate is used to store and carry liquids, food, and drinks [18]. PET is a thermoplastic that exhibits strong, glossy, and high impact resistance in addition to being exceptionally resistant to most solvents and weak acids and bases. Furthermore, compared to most other polymers, PET has a reduced gas permeability [19, 20]. Due to its excellent mechanical performance, thermal stability, nontoxicity, low processing energy requirements, and chemical resistance for a variety of applications, polyethylene terephthalate is one of the most widely used semi-crystalline thermoplastic polyesters in both industrial and everyday applications. The growing population, rising disposable income and spending power of consumers, and the flexible and convenient portability provided by beverage and food packaging are all predicted to contribute to an unexpected surge in the market for PET [20, 21].

Many different types of consumer goods, including water, soft drinks, alcohol, detergents, and cosmetics, are stored in these bottles or containers. Bottles for water and carbonated soft drinks are the two categories into which PET plastic bottles are divided [20]. PET resin-based bottles are easily recyclable and can be used again to create new products. The most popular plastic product on the market is PET bottles. PET is used to make the vast majority of soft drink bottles worldwide. PET is a popular material for bottles because it is inexpensive, widely available, lightweight, safe, and easily recyclable. It can also be easily moulded into a variety of designs [22].

1.2.5 Polyvinyl Chloride (PVC)

One of the polymers that is most commonly used worldwide is polyvinyl chloride, or PVC. Among the earliest thermoplastic polymers is PVC. Vinyl chloride (chloroethene), which is produced by chlorinating ethene, is polymerized to create PVC. Its gas barrier is strong, and while its moisture barrier is not as excellent, it can be made better by coating it with polyvinylidene chloride (PVDC), which also has the benefit of being easily heat sealable—a feature that is used in blister packs for prescription tablets. PVC that has not been plasticized is inherently brittle, and flexibility increases with increasing plasticizer content. Plasticizers improve processability by lowering the glass transition temperature, but they also reduce barrier qualities. As a result, a variety of grades are available to meet different end uses, ranging from hard bottles for greasy materials to extremely flexible films like cling film.

Due to its versatility, PVC is used in many different industrial, technical, and everyday applications. These include applications in the fields of construction, transportation, packaging, electronics, and medical. PVC is still used today for bath oils and was the very first substance that replaced glass bottles for cooking oil due to its superior resistance to grease and oil [5].

CHAPTER 1.3: LABELS OR IDENTIFICATION CODES FOR DIFFERENT TYPES OF PLASTICS

One of the materials most frequently utilized in a wide range of industries is plastic. Because plastics are an integral part of our daily lives, it is not unexpected that they produce a significant quantity of trash given their brief usage. This highlights how crucial and urgent it is to recycle plastic. Municipal Solid Waste (MSW) management is becoming a major concern for governments everywhere in the world [23, 24]. The two main processes in plastic recycling are the separation and sorting of plastic resins from MSW. Numerous plastic resins are used, most of which are incompatible with one another, which presents a special problem for the recycling process. Since contamination during the recycling of one type of plastic by another might result in significant processing issues, the separation of the various polymers by type is nearly always required [23, 25]. For instance, a resin batch would be rendered useless if there was more than 50 parts per million (ppm) of polyvinyl chloride in polyethylene terephthalate. Because polyvinyl chloride contains chlorine, it is unfavourable to the polyethylene terephthalate recycling process and reduces the quality of the final product, namely in terms of colour and viscosity [23, 26, 27]. Therefore, the resin identification and classification play a vital role in recycling process.

For easy classification during the classification process, in 1988, the Society of Plastics Industry adopted a standard marking code to assist consumers in identifying and sorting the primary types of plastic, as there are approximately 50 distinct types of plastic with hundreds of variations [28]. Nearly every plastic product has the universal recycling symbol, which is a

TABLE 1.1   Resins used in the manufacture of bottles

RESIN	RESIN IDENTIFICATION CODE	PRODUCT EXAMPLES	PROPERTIES	PRODUCT OBTAINED ON RECYCLING
Polyethylene Terephthalate (PET)	♳ PET	Water bottles, carbonated drinks bottles, food jars, oven-able films, clear bottles for storing dishwashing liquids, edible oils, etc.	Clarity, barrier to gas and moisture, heat resistant, reusable, tough	Fibres and drink bottles
High-Density Polyethylene (HDPE)	♴ PE-HD	Milk containers, juice bottles, packaging films, cosmetic bottles, crates, bins	Moisture and chemical resistant, tough	Toys, pens, bottles, tables, rope, etc.
Polyvinyl Chloride (PVC)	♵ PVC	Cleaning products bottle, chocolate trays, blister packaging, plumbing pipes, etc.	Resistant to grease, chemicals and oil, versatile	Pipe hoses, packaging materials, shoe laces, sewage pipe, etc.
Low-Density Polyethylene (LDPE)	♶ PE-LD	Thin and pliable materials like shopping bags, food containers, gloves, cosmetic tubes, etc.	Tough, easy to seal, moisture barrier, etc.	Plastic bricks, irrigation piping, dispensing bottles, etc.

Polypropylene (PP)	5 PP	Condiments bottle, bottle caps, ice cream containers, strapping tapes, etc.	Strong, tough, versatile, resistant to moisture, etc.	Automobile parts, tubs, industrial fibres, etc.
Polystyrene (PS)	6 PS	Display boxes, yoghurt and dairy product tubs, cake domes, etc.	Clear, versatile, insulated, easy to form objects, etc.	Pots, tubs, trays, decorative picture frames, etc.
Miscellaneous	7 OTHER	Baby bottles, CDs, number plates, storage containers, etc.	Properties depend on the resin composition	Car parts, pallets, etc.

triangle made up of three circling arrows. The type of plastic is indicated by the triangle's number. Six common kinds of plastic exist, along with a seventh miscellaneous category. Table 1.1 describes the resin identification codes, their use, and some common misconceptions about them [29].

CHAPTER 1.4: MANUFACTURING PROCESSES

Manufacturing techniques of plastic bottles

There are several key steps involved in the manufacture of the plastic bottles. They are outlined in Figure 1.1.

The details and the manufacturing techniques are discussed below:

Preforms or parisons:

A preform is a small plastic cast which has the exact shape of the bottle that is desired including the neck, complete with threads. The calculated amount of plastic is provided in the cast to give the specific wall thickness, colour and tensile strength and then hot pressurized air is blown into it to the bottle of desired shape and size.

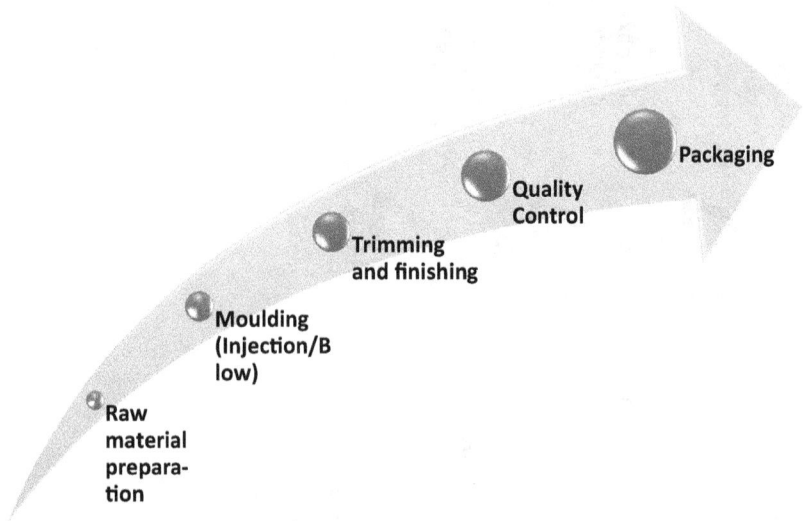

FIGURE 1.1   Key steps involved in the manufacturing of plastic bottles

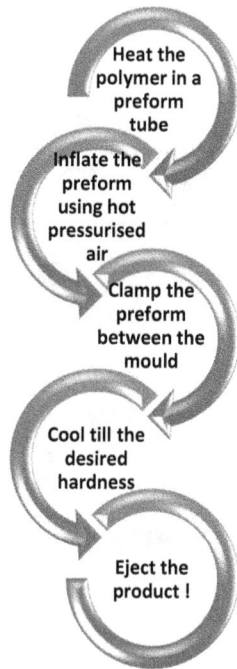

FIGURE 1.2   Basic steps in blow moulding techniques

A parison is a hollow plastic tube which is heated and blow moulded into the desired bottle shape while it is still warm. On cooling we get the desired shape and size of the bottle.

The blow moulding can be performed by different processes which we will discuss in the next segment, but broadly the process is a four step procedure as given the following flow diagram (Figure 1.2).

1.4.1 Extrusion Blow Moulding

This process is initiated by heating the material till it is workable and is then extruded into parison. A cooled mould is then used to clamp parison into position. A blast of hot pressurized air inflates the parison into the desired shape and size where it cools rapidly making contact with the colder metal (Figure 1.3).

The bottle so obtained will require trimming of the extra leftover material. This is observed as the pinch offline on the base of the bottle.

This trimming is usually done while the bottle is still in the mould else it will require an additional step along the production timeline.

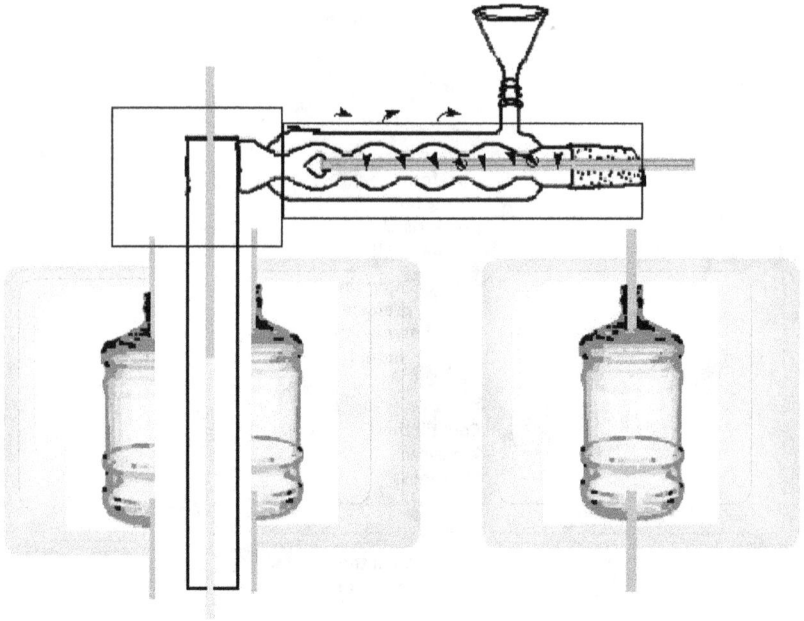

FIGURE 1.3 Extrusion blow moulding

1.4.2 Unique Characteristics

- The handles, labels moulded in and offset necks.
- Retain multi-layering of different materials into the same output bottle.

1.4.3 Plastics Commonly Used

- HDPE
- PVC
- PC
- PP

1.4.4 Injection Blow Moulding

This method is widespread among several plastic materials over a wide selection of sizes and shapes. Herein a preform of plastic material is heated and blown to obtain the desired bottle with high accuracy and thin-walled structure (Figure 1.4).

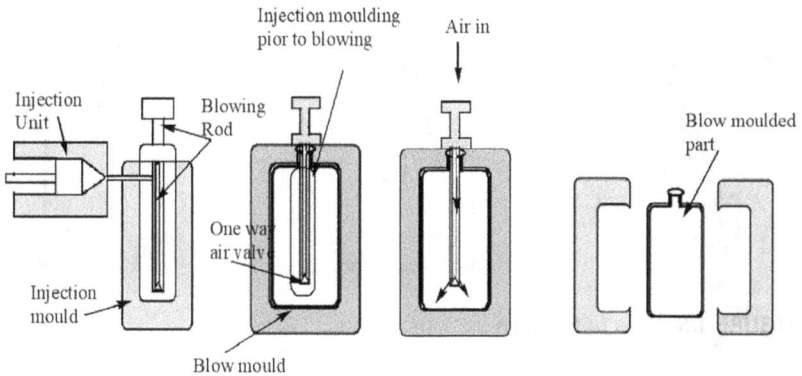

FIGURE 1.4 Injection blow moulding

It is imperative to note that the bottles obtained by this process are low in strength as polymer molecules are not stretched and oriented. It is therefore not suitable for carbonated beverages.

1.4.5 Plastics Commonly Used

- HDPE
- PET
- PP
- PVC
- LDPE

1.4.6 Injection Stretch Blow Moulding

This method is a slightly improvised version of standard injection blow moulding. In this process a metal rod is employed to stretch the blown plastic into its final shape. The process increases the overall wall strength of the bottle as the molecules get more firmly linked to one another giving the desired strength to the bottle. The slightly rigid molecular layout makes the bottle impermeable to gases thus making it ideal for carbonated drinks (Figure 1.5).

1.4.7 Plastics Commonly Used

- PET: The method gives a transparent shiny finish to the bottles and is used as a standard industry procedure across the globe.

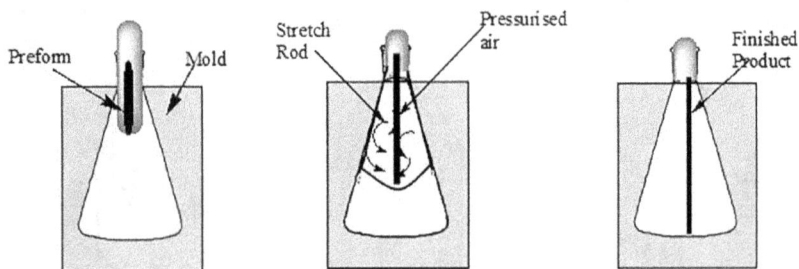

FIGURE 1.5   Injection stretch blow moulding

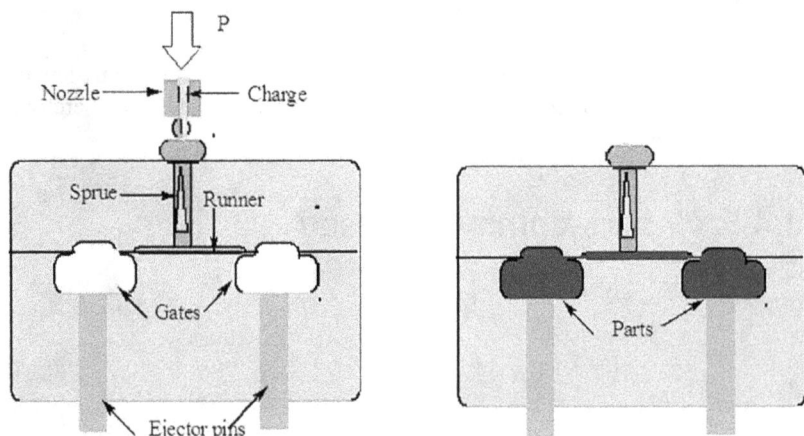

FIGURE 1.6   Injection moulding [30]

1.4.8 Injection Moulding

This method is used employed to manufacture hollow containers which are open on one side like tubs, caps and lids. They usually lack neck, threads and bottle contours. The material is injected into a cavity where pressure forces the resin to adapt to the mould body (Figure 1.6)

1.4.9 Plastics Commonly Used

- Thermoset plastics

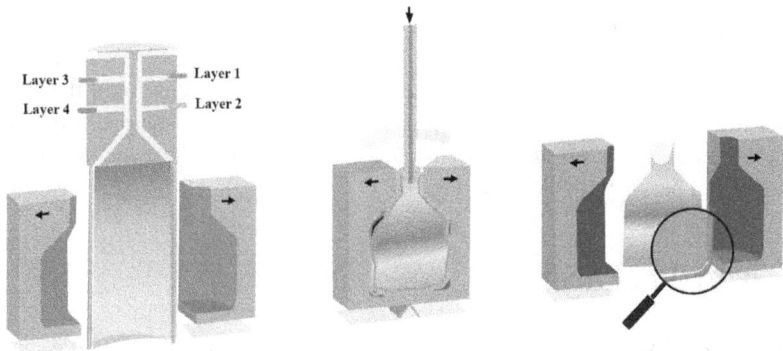

FIGURE 1.7   Co-extrusion

1.4.10 Co-Extrusion

Co-Extrusion is used to manufacture multi-layered bottles which employs the best possible combination of different polymers to be used for specific purpose. The method includes selection of suitable polymers with specific properties and then producing different layers of them for specialized needs. The aim of the layers is to increase strength and durability, fine tuning the permeability (mainly for oxygen, carbon dioxide, water vapour) and development of resistance towards chemicals so that the bottle material is not withered by the contents of the bottle (Figure 1.7).

1.4.11 Plastics Commonly Used

- EVOH, PVDC, PAN, PA: These polymers have low permeability coefficients and are used to create oxygen and carbon dioxide barriers. These polymers have higher resistance to chemicals.
- HDPE, PP: These polymers have good adhesive qualities as a result they are used as inter layer materials.

1.4.12 Trimming and Finishing

After moulding process is over, the excess plastic is trimmed off especially in the curved areas of the bottle. After trimming the bottles are inspected for any holes or deformations. Lables are then applied at times as adhesive labels or as in-mould labels.

1.4.13 Quality Control

There are different parameters that are applied for quality control of the bottles manufactured. Some important parameters are weight, volume, capacity and strength. They are also visually inspected for defects and another very important parameter is leak testing.

1.4.14 Packaging

After the above-mentioned steps are completed, the manufactured bottles are sorted and packed as per the count. They are then packed into boxes or onto pallets for transportation.

Plastic Bottles and the Planet

2

CHAPTER 2.1: LIMITATION OF PLASTIC BOTTLES IN PACKAGING

Although it's true that "plastic is the greatest discovery of the millennium," it's also true that plastic garbage has spread over the world since much of it degrades slowly. Specifically, the majority of developing nations lack access to innovative technology as well as appropriate laws and policies governing the manufacture, use, and disposal of plastics. By the end of 2015, it's predicted that 5,800 metric tonnes of improperly managed plastic garbage had been dumped into the environment worldwide, with a fair share coming from the drinks and beverage industry. Because of their large molecular weight, intricate three-dimensional structure, hydrophobicity, stability, and resistance to biodegradation, these plastics containers endure in the environment for decades [31, 32].

Plastics undergo a variety of weathering processes, such as hydrolysis, photodegradation, thermal oxidation, biodegradation, and fragmentation. Both biotic (microorganism-mediated biodegradation) and abiotic factors (photodegradation from UV radiation or weathered disintegration from wind and wave action) disintegration processes are applied to plastic waste in plastic bottles. These procedures produce what are known as microplastics (MPs), which are tiny pieces of plastic. MPs are classified as particles less than 5 mm in length, whereas nano plastics (NPs) were previously used to describe plastic particles less than 1 μm [33]. A few of the limitations of using plastic bottles have been mentioned below.

2.1.1 Very Slow Degradation Rate

In landfills, plastic products might take up to a millennium to break down. Plastic bottles take 450 years to dissolve, whereas plastic bags break down in 10–20 years when used regularly. Thermoplastics typically takes a landfill between 70 and 400 years to break down. Bacteria that eat garbage and break it down into simple materials are essential to both processes. It is challenging for bacteria to take in PET since it is composed of substances that they cannot eat. Plastic's chemical structure is broken down by ultraviolet (UV) radiation from the sun, resulting in progressively smaller pieces over time. It is not unusual for this procedure to take years because compounds, like PET, are quite resistant. The disposal of plastic bottles has been discussed in sections ahead.

2.1.2 Effect on Human Health

Plastics have certainly contributed significantly to the advancement of mankind's civilization, but the environmental proliferation of plastic trash, including macro-, micro-, and nanoplastics, as well as its incorporation in biological systems, are currently major causes for concern. Many health concerns, including as obesity, diabetes, thyroid dysfunction, and reproductive disorders, have been connected to plastic pollution. As an illustration, research has shown that nanoplastics negatively impact the variety and make-up of the microbial communities found in the human gut. Considering that there is a close association between the neural networks in the gut and the brain, this could have a negative impact on the immunological, endocrine, and neurological systems.

A litre of water in a plastic bottle was discovered to contain, on average, 240,000 identifiable plastic bits, which is 10–100 times more than what was previously estimated. Mason et al. [34] found an average of 325 pieces of microplastics in a litre of bottled water. Because they are so tiny, micro- and nano plastics can enter the circulation directly after slipping past the lungs and intestines protecting layers. The brain and heart are among the organs to which the pieces may subsequently travel. Many studies have also connected obesity, insulin resistance, and thyroid and endocrine system disruption to micro- and nanoparticles. Additionally, these particles are also able to pass past the placenta and into the bodies of foetuses. It is known that micro- and nanoplastics are genotoxic to DNA. It has been shown that damage can occur if the plastic matter is small enough to get through the nuclear membrane that surrounds the DNA. This can lead to lesions or damage to the DNA structure, which, if left untreated, can result in mutagenic processes that are thought to be involved in the carcinogenesis of cells. Furthermore, it was discovered that the form, functional groups, and chemical makeup of the plastic waste all affect the kind and degree of DNA damage (Figure 2.1).

Bottles made of plastic also pose a risk to human health because of their additives (like plasticizers) or monomeric building blocks (like bisphenol A) or a combination of the two (like antimicrobial polycarbonate). Plastics release various harmful substances into the environment. We focus on those that are of primary concern when it comes to plastics, like phthalates and bisphenol A.

The monomeric component of polycarbonate plastics, bisphenol A (BPA), is the compound that is most well-known. First created in 1891, it is commonly added to other plastics, like polyvinyl chloride (PVC), as an additive. In 2003, 2.2 million metric tons of BPA were produced annually worldwide. This bulk comes into contact with foodstuff to a considerable extent. Certain monomers remain unbound following BPA polymerization, which allows BPA atoms to eventually leak out of food and drink containers into beverages. BPA is xenoestrogen, sometimes referred to as environmental oestrogen, which are artificial or naturally occurring substances that replicate the physiological effects of oestrogen. BPA has been connected in numerous studies to a wide range of health problems, such as immune system changes, quick puberty, ovarian chromosomal damage, obesity, cardiovascular disease, type 2 diabetes, and sperm production reduction. Furthermore, several studies have indicated that BPA increases

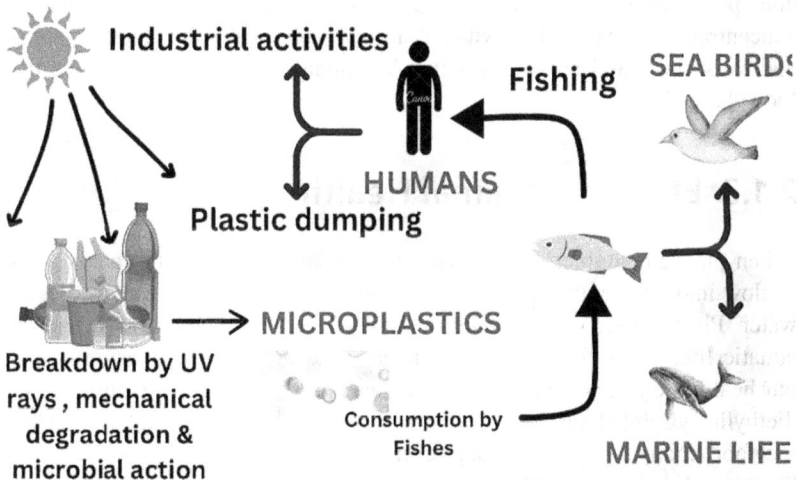

FIGURE 2.1 Pathway of microplastics in the living system

the risk of metabolic disorders, pain, prostate and breast cancer, among other health issues. Additionally, BPA has been connected to a number of detrimental health outcomes for women, including obesity, polycystic ovarian syndrome, recurrent miscarriages, endometrial hyperplasia, and sterility.

It has long been known that components of plastic packaging migrate or interact chemically with foods high in fat; one such interaction is the transfer of antioxidants from the packaging made of plastic into the material of consumption, which can occasionally attach to the item's surface. There may be a health concern from this kind of transfer of packaging additives from the package material to the product inside. In addition, endocrine disruptors can be found in PET, a major plastic used in the beverage and food sectors. These disruptors can seep into the consumables the plastic packaging holds. Phthalates, which have the potential to cause endocrine disruption, have been observed to seep from PET packaging into different food items when there is water present, even at room temperature [35].

Numerous studies have shown that the chemicals used in the production of plastic are detrimental to human and animal health, and that these microplastics and nanoplastics can be extremely dangerous once they enter the bloodstream. However, little research has been done to date to determine exactly what happens as a result of this. The hazardous chemicals used in the production of plastics bottles, such as bisphenols, phthalates, dioxins, organic contaminants, and heavy metals, can be carried by nano plastics even if they don't pose a threat to human health. These chemicals can be harmful in high concentrations and can affect vital organs like the kidneys, liver, heart, reproductive system, and nervous system. Additionally, they may amass via the food chain [36].

2.1.3 Effect on Animal Health

When plastic containers, like plastic bottles, are dumped into coastal seas or flow into watercourses, they contaminate the oceans and other bodies of water. Plastic pollution is having a catastrophic and quickening effect on aquatic life. The effects of plastic trash, in general, on the ecosystem and our health are just now starting to become clear. Because of the emission of diethylhexyl phthalate, lead, mercury, and cadmium, this pollution reaches the food chain and poses a long-term risk of cancer to fish, animals, and humans. Microplastic waste that floats on the water surface typically contaminates oceans. The majority of marine turtles are harmed by plastic pollution, which also affects whales by accumulating in their stomachs and causing oesophageal obstruction in certain jellyfish species. The minuscule plastic fragments beneath the ocean's surface are also eaten by small fish. Plastics are inadvertently consumed by tuna, swordfish, and lantern fish and end up in the aquatic food chain.

Along with harming animals, plastic pollution also affects birds, such as seabirds, whose digestive system obstructions result in tissue damage from the harmful compounds known as polychlorinated biphenyls (PCBs). Even birds that have never visited the sea can be affected by marine plastic pollution due to their eating habits. Along with other plastic waste, such as styrofoam mixed with their food, the plastic fragments were discovered whole inside the birds' proventriculus and gizzards [37].

CHAPTER 2.2: ENVIRONMENTAL IMPACT OF PLASTIC BOTTLES PRODUCTION AND DISPOSAL

Approximately 400 billion plastic water bottles are produced annually and over a million are bought every minute worldwide. By 2021, this number is expected to have grown exponentially to around 600 billion. It is astounding to find out that 91% of plastic bottles are made for single use (also known as single-use plastics or SUP), and 8.3 billion metric tons of plastic pollution have been produced since plastic became widely utilized about 60 years ago due to mass manufacture. Our lives now consist of so many plastic bottles that we almost ever attach any significance to them. In actuality, they cost millions of dollars annually and have unimaginable effects on the environment.

Plastic bottles cannot be disposed of safely, and it has a negative impact on the environment all throughout its lifecycle—during production, use, and disposal. The release of hazardous synthetic compounds during the assembly process is another major source of the detrimental effects of plastics. A wide range of synthetic compounds that are carcinogenic, neurotoxic, and hormone-disruptive are common additives and waste products of plastic production, and they inevitably find their way into our environment through contamination of the air, water, and land.

In landfills, plastic bottles take about 450 years to completely decompose, and in natural ecosystems, it takes even longer. Over time, plastic bottles accumulate in our oceans, creating vast regions of floating plastic waste, such as the Great Pacific Garbage Patch. Captain Charles Moore found a vast expanse of contaminated plastic waste in the North Pacific Ocean in 1997. This area is known as a gyre. The projected contaminated area grew to 10 million square kilometres by 2005. It was found that 90% of this trash was plastic, and 80% of it originated from land, including building waste. According to reports, there are six such gyres filled with plastic debris throughout the oceans of the world. The Great Pacific Garbage Patch is estimated to contain 2 million tonnes of plastic waste.

Plastic bottles eventually break down into tiny bits on land and in the ocean. Polychlorinated biphenyls (PCBs), among other dangerous chemicals, are present in these components. Toxins are released in high amounts that can kill microorganisms in soil and water when plastic particles finally break down. One of the PCBs that has been examined the most is bisphenol A (BPA), which is frequently generated when plastic water bottles break down. According to studies, BPA functions in the human body similarly to oestrogen. An increased chance of developing some chronic illnesses, such as diabetes, cancer, and asthma, has been connected to this. Plastics, primarily in the form of micro- and nano plastics, are easily absorbed by the body and can lead to serious health problems like asthma-like syndrome, chronic bronchitis, pneumothorax, coughing, shortness of breath, phlegm, wheezing, frequent fevers, and generalized aches.

THE GREAT PACIFIC GARBAGE PATCH

Hawaii and California are separated by the Pacific Ocean, which contains the Great Pacific Garbage Patch (GPGP). The GPGP is constantly shifting in terms of both position and shape, though, as a result of shifting winds and ocean trends. The Eastern Garbage Patch extends from California to Hawaii, whereas the Western Garbage Patch extends from Hawaii to Japan. These two regions make up the GPGP. Even though it's known as a "plastic island," people occasionally confuse the GPGP for a real island. But because of how close the contaminants are to one another, it's merely a patch of floating rubbish on the ocean's surface that gives the impression of being an island.

An estimated 1,170–2,450 kg of plastic are washed into the ocean annually by rivers. According to a research study, there are between 1.1 and 3.6 trillion plastic particles in the Great Pacific Garbage Patch. It is estimated that each person on the planet uses about 200 pieces of plastic.

Irrespective of its size or location, it is crystal clear that the Pacific Garbage Patch isn't welcomed in our ocean, and we need to know more about it in order to appropriately address it.

The greatest method to lessen the harmful effects of plastic bottles is to cease using them, although recycling plastic and purchasing recycled plastic goods are also helpful options.

Glass, aluminium, or stainless-steel reusable bottles are a few of the most effective substitutes. The demand for reusable water bottles is enormous. Reusable bottles are available with a wide range of characteristics to accommodate practically everyone's demands. To drink tap water safely, metal pitchers with integrated water filters are a great option. The handling and disposal of plastics after use is crucial in mitigating the plastic problem. Recycling plastics is a crucial component of trash reduction, as is searching for environmentally friendly substitutes.

CHAPTER 2.3: RECYCLING CHALLENGES

Although the production of plastic bottles has increased dramatically over the past 70 years, it is just as essential for us to realize that the majority of these bottles are non-biodegradable and may take decades to break down. Thus, the only viable solution is to recycle these bottles. But even during recycling, these plastic bottles provide substantial obstacles to efficient recycling. Plastic trash poses particular challenges, in contrast to metals, which are generally easier to recycle. As previously mentioned, many types of polymers, such as LDPE, HDPE, PP, PVC, PS, and PET, are used in the production of plastic bottles. Polymer recycling is further complicated by the combination of various polymers and possible contamination from metal, paper, ink, pigment, and beverage types. The lack of facilities for collecting and sorting plastic garbage, the difficulty of efficiently sorting various sorts of plastic, and the high expense of gathering and handling plastic waste all make recycling plastic bottles more challenging. Plastic recycling carries increased expenses, which discourages both investors and producers. The acceptance of plastic recycling activities is hampered by this issue, which poses a significant hurdle.

Further, temperature, oxygen content, and UV radiation are three environmental elements that have a significant impact on the rates of polymer breakdown. The quality of recovered plastic material get generally declined as a result of these environmental factors, which also present obstacles to recycling activities. Recycled plastics frequently show poorer quality than their non-recycled counterparts, with colour fluctuation and reduced strength being two common characteristics. They can be less desirable to manufacturers and more difficult to incorporate into new goods, which is why these quality difficulties are obstacles to their use in the manufacturing sector.

For instance, PET resin is utilized in nearly every kind of plastic bottles sold worldwide. This material can be easily recycled into bottles, however keeping the bottle's quality intact during the recycling process is difficult and may cause deterioration and trash to be produced. Furthermore, bottle caps—which are frequently made of polyolefins—have outstanding recycling qualities. However, due of the additives, including colorants, their recyclability is frequently low. In order to overcome these constraints and improve the appeal and suitability of recycled materials across a range of industries, more focus and research are required due to the deteriorated quality of recycled plastics.

CHAPTER 2.4: ADVANCEMENTS IN RECYCLING CHALLENGES

When a plastic bottle reaches the end of its useful life, there are few options for what to do with it, despite the fact that recycling plastic bottles has significant economic and environmental benefits. Sorting these bottles is necessary before recycling, and it takes money and effort. In addition to being energy-intensive, recycling frequently yields low-quality polymers, as was previously mentioned. Mechanical recycling is the only widely employed technology for the large-scale disposal of plastic solid waste. Washing away any organic residue is the first step in the primary procedures. After that, the polymer is shredded, melted, and remoulded. To make a material with the proper properties for production, it is often blended with virgin plastic of the same kind.

Mechanical recycling technologies have certain restrictions since the chemical composition, mechanical behaviour, and thermal qualities of different types of plastic affect how they react to the process. One cannot process polymers manually, including temperature-sensitive plastics, composites, and plastics that don't flow at high temperatures (like thermosets). Because of this, only two types of plastic—PET and polyethylenes—are recovered and recycled using mechanical processes. These two types of plastic account for 9% and 37% of the total amount of plastic generated annually, respectively. The remaining plastic solid waste is either not recovered at all or is recovered to a level that is less than 1% of output.

Mechanical method is not suitable for recycling many polymeric materials. According to recent research, it is possible to create chemical recycling methods that use less energy, to improve the compatibility of mixed plastic wastes so that sorting is not necessary, and to apply recycling technologies to plastic bottles made up of polymers that are not normally recyclable. Chemical recycling refers to current technologies that go beyond mechanical recycling, such as pyrolysis (thermolysis), which uses catalysts to selectively create gases, fuels, or waxes.

The hydrolysis and pyrolysis of used plastic bottles are common chemical processes. After that, the product is utilized as a feedstock to make polymers and fuels. The last recycling method involves burning the polymer to recover energy. The polymer is burned during this process, and some energy is recovered as heat. When no further value-added application is possible, this procedure is typically used as a "last resort." Many plastic bottles that are incinerated also release harmful gases and release toxic byproducts, which can have negative ecological effects and increase the expense of collecting and remediating hazardous garbage. Numerous studies are being conducted to examine the most efficient ways to recycle plastics using a combination of chemical and mechanical processes [38]. Table 2.1 explains the recycling techniques for different polymers used in making bottles.

TABLE 2.1   Recycling information of resins used in plastic bottles

RECYCLING CODE	RECYCLING INFORMATION
△ 1 PET	PET bottles can be recycled mechanically by being cleaned, shred, and melted to create new plastic resin. High-grade recycled PET resin may be derived by chemical recycling, which is also researched.
△ 2 PE-HD	It is hard to melt and shred because of its rigidity. HDPE materials are gathered, segregated, thoroughly cleaned to eliminate impurities, shred, melted, and subsequently moulded into new items, including toys, crates, and bottles.
△ 3 PVC	After being gathered, sorted, and cleaned, it is finely chopped into tiny fragments, similar to flakes and granules. Subsequently, chlorine is eliminated chemically and the area is cleaned once more in order to create fresh PVC materials.
△ 4 PE-LD	The procedure is essentially the same as with HDPE, but because of its great flexibility, which tangles with particles easily, sorting becomes challenging. Also, LDPE products are more likely to become contaminated.
△ 5 PP	The shredded PP is first collected, sorted, cleaned, and then shred. Next, it is melted at an extreme temperature and sent through a device to generate consistent plastic resin.

(Continued)

TABLE 2.1   (Continued)

RECYCLING CODE	RECYCLING INFORMATION
6 PS	In order to decrease its volume, PS foam in particular is compressed or densified after being collected, sorted, cleaned, and shred. The foam is heated and squeezed into denser blocks for this purpose. After that, a machine is used to extrude it, producing pellets of homogenous plastic resin.
7 OTHER	Depending on a number of variables, including flexibility and rigidity, recycling different plastics might be straightforward or complex.

2.4.1 Incineration

The process of burning garbage in oxygen, referred to in chemistry as "complete combustion," is what is meant by the term "waste incineration." Carbon dioxide and water molecules are released into the atmosphere as a result of this type of burning. The residue that is left over after burning is composed of ash, several volatile chemicals, and a little amount of hydrochloric acid. Certain waste plastics are resistant to heat, oxygen, and explosives, thus they cannot all be burned efficiently.

There are some benefits to burning plastic bottles as a trash management technique. These include a decrease in the quantity of waste the ecosystem produces, an increase in the quantity of heat and power that can be used for a range of purposes, a reduction in the quantity of pollution released into the surrounding environment, financial savings from the elimination of potentially dangerous chemicals and germs, and a decrease in the quantity of waste released into the atmosphere. It can also be used in any climate and at any point of year because it prevents methane gas from being formed. As a chemical process, incineration has benefits and drawbacks much like any other biological or scientific endeavours.

Among the many disadvantages of the incineration process is its expensive setup when compared to other trash disposal options. It contaminates the environment and endangers human and environmental health. It causes the discharge of waste ash, which is dangerous for people and the environment. Air pollution is further exacerbated by the burning of plastic bottles in open spaces. Hazardous materials such dioxins, furans, mercury, and polychlorinated are released into the atmosphere when municipal solid trash, which normally contains 12% plastics, is burned. Burning plastic garbage, such as bottles, damages the nervous system and increases the possibility of cardiovascular disease, emphysema, and asthma attacks. It also aggravates pre-existing respiratory disorders and can produce irritation to the skin, nausea, and headaches. Consequently, scientists and environmentalists should concentrate their efforts right now on making a sustainable change that will lead to a better and cleaner environment down the road.

2.4.2 Pyrolysis

Pyrolysis is the process of converting gases and fatty oils into hydrocarbons and crude petrochemicals. Hydrocarbons can also be produced through pyrolysis. With its application, it is also conceivable to recover crude petrochemicals and produce sustainable energy from waste plastics. Based on the quantity of heat energy required for breaking down plastic connections, the pyrolysis process can be divided into three main groups: those based on high temperature, medium temperature, and low temperature. The range of temperatures that are necessary to cause the plastic's structure to break down defines what constitutes a medium and high temperature. The following ranges of temperatures are associated with the pyrolysis states: under 600°C, between 600°C and 800°C, and above 800°C. The products that result from the breakdown of polymers depend on a number of factors, such as the type of reactor utilized, the length of time the plastics spend in the reactor, the kind of plastics, the configuration of the feeding and condensation lines, and the applied temperature. Crude oil-derived plastics are broken down into their monomeric units and other usable components, like plasters and additives, which are then classified as the first form of petroleum recycling cuts and petrochemicals. Thermal radiation or catalytic chemical diagnostic techniques are used to effect this change. These chemical treatment methods were created for the waste management sector to enable the proper accumulation of plastic wastes and their effective management. The scientific method presented here is an efficient way to control waste, even if it could require a significant financial outlay.

2.4.3 Degradation of Plastic Bottles

The growing demands on plastics require the development of an efficient waste plastic processing system in order to maintain a sustainable equilibrium. It is the biodegradation process. Biodegradation is an efficient method that is profitable and useful for the processing of waste plastics. The ability of a sizable number of microorganisms to break down plastic polymers is advantageous since it might be used to combat problems caused by the everyday accumulation of more plastic waste. Enzymes that break down polymeric polymers into smaller bits are produced by a variety of microorganisms. Some of these enzymes are found extracellularly, while others are located intracellularly.

Abiotic stimuli and a number of bacteria work together during the biodegradation process to break down polymers into smaller molecules. The subsequent phase in this process will be depolymerization. Enzymes and free radicals are released by bacteria to form biofilms, which help the microbes break the polymer chains gradually and aid in the process of depolymerization. Biodegradation, or the modification of plastic polymers, is a process that certain bacteria can induce on the surface of plastic. Over the past few decades, plastic pollution has become a significant issue due to the limited capacity of the material to be recycled and biodegraded.

2.4.4 The Use of Microbes in Plastic Degradation

There has been a significant advancement in the utilization of microorganisms in the biosorption of polymers. Since polymeric molecules are the building blocks of plastics, they might be the only source of carbon for the microbial population. The creation of a zone referred to as the plastic sphere may result from microbes' capacity to produce biofilms on the surfaces of contaminants. Microbes in this area can collaborate to produce acids or other enzymes that aid in the breakdown of plastics. A plastic sphere's potential microbe composition depends on a number of factors, including the kind of polymer, its size, surface characteristics, and other external factors. Biodegradation is the next stage of the process that occurs when physical and chemical breakdown impair the structure of polymers.

It is possible for microorganisms to break down the plastic surface because they produce extracellular polymeric substances (EPS). EPS is composed of three main components: nucleic acids, proteins, and polysaccharides. The entry of EPS through the holes causes the plastic's surface pores to expand. Because plastic polymers are more prone to bacterial and microbial degradation, holes may emerge in them, and the materials' physical degradation may accelerate.

The polymerization of the constituent parts of a polymer is accomplished by enzymes referred to as depolymerase. Monomers, amides, and oligosaccharides that may be created as a result of this process are less complex than polymers. Additional processing will be applied based on how many oxygen molecules are involved in the process of metabolism. After these components break down aerobically, biomass from microbes, carbon dioxide, and water vapor are produced. A microbial population, carbon dioxide, water, and either hydrogen sulphide or methane will be the results of anaerobic breakdown, on the other hand. Plastic garbage breaks down because of the vital responsibilities that microbes play in producing extracellular and intracellular depolymerase enzymes that are then discharged into the environment.

2.4.5 Transforming Plastic Waste into Fuels through Recycling

Enzyme activity causes a specific component of a polymer to eventually break down into simpler molecules, a biological process known as biodegradation of polymers. Plastics frequently biodegrade in two different ways: anaerobically in landfill and silt and aerobically in soil and compost. Carbon dioxide and water are produced during aerobic biodegradation, whereas carbon dioxide, water, and methane are produced during anaerobic biodegradation. When plastic waste is exposed to the biodegradation process, nanoplastics may be created. Gaining a thorough understanding of the interactions that occur between plastics and their surroundings in their natural environments is of paramount relevance.

The hydrolysis of polyethylene, polypropylene, and PET is accomplished by photo, thermal, and biological degradation processes. The breakdown of the three separate polymers occurs at varying rates and in distinctive ways. In normal circumstances, the effects of thermal and light deterioration are similar. Polyethylene's infrared spectra show bands representing ketones, esters, acids, and other compounds that become sharper as a result of photodegradation. With the additional benefit that polypropylene is less prone to photodeterioration, the same thing may be said about it. The photo-oxidation of PET produces hydro peroxide species by oxidizing the CH_2 groups near the ester linkages. Consequently, a variety of pathways lead to the production of photoproducts from these hydro peroxide molecules. Each of the three different polymers interacts with microorganisms and develops biofilms in a different way. The carbonyl indices will normally decrease as a result of the biodegradation process if the sample has previously been photo-degraded by exposure to UV radiation. The local environmental conditions, which are frequently a combination of the elements that are mimicked in laboratories, have a significant influence on the rate of plastic degradation.

2.4.6 Chemical Recycling

The best way to stop the plastic bottle epidemic is to find natural alternatives, but the next challenge is to make plastics biodegradable on a wide scale in a way that is both sustainable and economical. The processes involved in mechanical recycling include sorting, melting, and remoulding polymers into lower-grade plastic products. The drawbacks of this approach are that polymers lose some of their performance attributes with each recycling cycle. Polymers can be recycled into useful resources by causing the plastic bottles to break down at the molecular level by chemical recycling. Pyrolysis is a process of thermal breakdown in which some polymers, such as polyolefin, are burned at high temperatures to produce waxes and fuels [39].

CHAPTER 2.5: RECYCLING THROUGH VARIOUS CASE STUDIES

The global effort to reduce plastic waste and safeguard the environment now includes recycling plastic bottles as a critical component. As awareness of plastic pollution's detrimental effects on ecosystems and public health has grown, numerous businesses, groups, and campaigns have created creative ways to improve recycling procedures and advance the circular economy. In addition to highlighting various tactics, tools, and cooperative models that support more effective waste management and sustainable practices, this case study introduction gives a summary of noteworthy initiatives and achievements in the recycling of plastic bottles.

By looking through these case studies, we may learn how forward-thinking businesses and projects address the problems associated with plastic waste by using cutting-edge recycling technologies, creative packaging ideas, and successful collaborations. Each case study showcases distinct strategies and accomplishments in enhancing the recycling of plastic bottles, ranging from multinational enterprises executing expansive recycling systems to grassroots organizations spearheading community-based solutions. The knowledge gained from these cases not only shows the progress that has been done, but it also offers insightful guidance and motivation for future initiatives to lessen plastic pollution and promote a more sustainable future.

2.5.1 Coca-Cola's "World without Waste" Initiative

Through this program, Coca-Cola hopes to gather and recycle enough bottles and cans to cover its global sales by 2030. To increase the sustainability of its packaging, the company has made significant investments in mechanical and chemical waste recycling technology.

2.5.1.1 Significant Components

- Collection and Sorting: To handle different forms of plastic, there is an expanded infrastructure and cutting-edge sorting technology.
- Mechanical recycling: Involves washing, cutting, and reusing PET bottles to create new goods.
- Chemical Recycling: The process of depolymerizing PET into monomers to produce high-grade recycled PET.
- Recycled Content: A pledge to use 50% recycled materials by 2030 in PET bottles.

2.5.1.2 Outcomes

- It resulted in a greater incorporation of recycled resources into new goods.
- This initiative has also increased recycling rates worldwide and improved sustainability initiatives.
- It has led in the creation of collaborations and local recycling initiatives.

2.5.2 The "Plastic Waste-Free World" Initiative from Unilever

Unilever has initiated a comprehensive strategy aimed at curbing plastic waste and improving recycling practices across its business. The company specializes on better recycling technology and creative packaging solutions.

2.5.2.1 Significant Components

- Packaging innovations include the creation of materials that are biodegradable, recyclable, and reusable.
- Recycling Partnerships: Working together to enhance the infrastructure for recycling with local authorities and groups.

- Engagement of Consumers: Initiatives to inform customers about appropriate recycling and disposal procedures.

2.5.2.2 Outcomes

- New packaging designs with more recycled material were introduced as a result of it.
- Plastic trash has decreased and recycling rates have increased as a result of this.
- In important markets, the program has also improved the infrastructure for recycling.

2.5.3 TerraCycle's Loop Program

Reusing packaging is emphasized in TerraCycle's Loop concept, a circular shopping approach. Customers buy goods in robust, reusable containers that are gathered, cleaned, and then filled again.

2.5.3.1 Significant Components

- Reusable Packaging: Reusable containers are gathered for cleaning and replenishment and are made to last for several usage.
- Partnerships: Working together with well-known companies and merchants to provide goods in reusable packaging.
- Customer Involvement: Put your efforts into developing a system that promotes customer involvement in waste reduction.

2.5.3.2 Outcomes

- It took the lead in using reusable packaging to reduce the waste from single-use plastics.
- It improved the creation of a scalable paradigm for environmentally friendly packaging options.
- As a result, people became more conscious of and involved in sustainable consumption behaviours.

2.5.4 The Initiative for Plastic Banks

Using a circular economy to handle plastic trash, the Plastic Bank program seeks to reduce plastic pollution in the ocean. Supporting recycling infrastructure in underdeveloped nations, the group offers incentives for collecting plastic waste.

2.5.4.1 Significant Components

- Collection Incentives: Offering monetary rewards as well as other advantages to encourage the collection of plastic garbage.
- Centres for Recycling: Setting up centres for recycling to transform gathered plastics into new goods.
- Collaborations: Bringing recycled plastic into new goods by collaborating with companies and communities.

2.5.4.2 Outcomes

- In coastal communities, it led to a notable decrease in plastic trash.
- It resulted in the development of a circular economy, which boosts regional economies and the environment.
- Furthermore, the project enhanced waste management techniques and raised recycling rates.

2.5.5 PureCycle Technologies

PureCycle Technologies specializes in a ground-breaking procedure known as "Polypropylene Recycling," which purifies and recycles polypropylene (PP) plastics using a patented solvent-based technique.

2.5.5.1 Significant Components

- Solvent-Based Purification: An innovative method that yields high-grade recycled polypropylene by eliminating impurities and colours from the material.
- Scalable Technology: Establishing expansive facilities for the processing and recycling of waste polypropylene.
- Partnership: Collaborations with well-known companies to incorporate recycled polypropylene (PP) into new goods.

2.5.5.2 Outcomes

- High-purity recycled polypropylene that is appropriate for a variety of applications was produced as a result.
- It had a smaller negative environmental impact and decreased the requirement for virgin polypropylene manufacture.
- The program assisted in expanding the recycling capacity for a widely used yet difficult-to-recycle plastic.

Biopolymers
An Emerging Alternative

3

CHAPTER 3.1: POTENTIAL APPLICATION OF BIOPOLYMERS IN PACKAGING OF BOTTLES

Even while we can't entirely avoid using plastic bottles, we should look into more practical and superior alternatives. Many sustainable and recyclable polymers have been created in laboratories thanks to advances in science and technology; nonetheless, the cost and process of producing these plastics remain an issue. The cost of using non-plastic substitutes in consumer goods and packaging is four times higher than that of plastics. The cost of manufacturing biodegradable plastics is twice that of regular plastics. However, given that plastic takes more than 400 years to decompose, it is a significant element of the current geological epoch, and has produced the plastisphere, a new microbial habitat, it is imperative that we find an alternative to save the planet. Modern polymers and bottle packaging have a number of substitutes. We go over a few of these plastic varieties below.

Biodegradable plastics are plastic polymers that, in specific environmental circumstances and with the assistance of living creatures, break down into entirely natural elements such as water, carbon dioxide, and compost. Biodegradable plastic bottles can be made from fossil fuels, natural resources, or biobased materials.

A plastic called **bioplastic** is made from natural raw ingredients, like corn, sugar, starch, cellulose, potatoes, cereals, molasses, and soybean oil. As

DOI: 10.1201/9781003541103-3

it is made in a sustainable manner, bottles made of bioplastic could either be or not be biodegradable.

Synthetic polymers made from living things are called **biopolymers**. Covalent bonds between numerous monomeric molecules produce a large structure. The polymers that occur naturally that are used to produce plastics are referred to as biopolymers. Most of these polymers originate from regenerative sources, such as plants and animals. Biopolymers are organic compounds that are present in naturally occurring sources. The name "biopolymer" originates from the Greek concepts "bio" and "polymer," which denote environment and living things, respectively. Large macromolecules called biopolymers are made up of several units that repeat. The biopolymers are useful for a range of applications since it has been found that they are both biocompatible and biodegradable. These applications include dressing materials in the food and beverage and pharmaceutical industries, edible films, emulsions, drug transport materials, and medical implants.

The following are the reasons biopolymers attracted a lot of interest in the food and beverage industries:

i. Due to their great diversity and minimal mechanical property modifications, biopolymers can be used in a wide range of industries for packaging and preservation.

ii. Biopolymers have the potential to be renewable, carbon neutral, and sustainable. The agricultural non-food crops are their origins. They are therefore a trustworthy supply in these fields due to their environmental soundness.

iii. Within six months, 90% of biopolymers decompose into compost, and some of them are also biodegradable. Therefore, if they are deployed, they can stop pollution from synthetic polymers that harm the environment.

Plant or animal biomass, specifically in the form of polysaccharides and protein, is the direct source of biopolymers. All living things have polysaccharides during their growth cycle, and every bodily cell has protein. They are both utilized to create materials that decompose naturally. Direct plant or animal derived materials make up biopolymers. Packaging materials can be produced using biopolymers such as polysaccharides, lipids, proteins, or occasionally a combination of these. It has a great chance to replace the synthetic polymers that are already on the market.

The biopolymer packaging industry was anticipated to grow at a rate of roughly 22.68% between 2014 and 2019, based on data from the global market projection. The most common application of biopolymers, as in the global scenario, is in packaging. Furthermore, it is anticipated that ongoing research and novel discoveries in the field of biopolymers will propel the market throughout the forecast period. Due to its enhanced time-demanding properties, biopolymer packaging has recently become widely used in the food and beverage industries.

CHAPTER 3.2: TYPES OF BIOPOLYMERS IN BOTTLE PACKAGING

Polymers can be elastic or rigid, permeable or impermeable, hydrophilic or hydrophobic, and have a wide range of other material qualities. The repeating monomer building blocks of polymers, or their structure, dictate these features. We refer to the polymeric substance as plastics once it has been processed and shaped—usually with heat—into its final, economically relevant shape. There are two types of plastics: thermosets and thermoplastics. The former are tougher than the latter and their shape is mostly unaffected by temperature. The former are used in car tires and epoxies. The majority of plastics are thermoplastics, which are made up of linear chains of polymers which permit thermal reshaping, like those employed in bottles and textiles.

While fossil resources are used to make the majority of commercial plastics, renewable resources can also be used to create these materials, such as bioplastics or biopolymers. Here, the monomers are manufactured or extracted from biomass components (plant sugars, for example), and they are then polymerized to create either new polymers like polyhydroxyalkanoates (PHAs) or cellulose, or they can be used as a direct substitute for existing plastics like polyethylene (PE) and cellulose. Some biobased plastics and biodegradable plastics have been discussed ahead in depth [40].

3.2.1 Poly Lactic Acid (PLA)

Biopolymers for liquid food packaging ought to possess comparable resistance, barrier, inertia, long-term stability, and clarity to traditional plastics like polyethylene terephthalate (PET). These specifications for liquid food packing can currently only be met by PLA and its composites. Water, carbonated beverages such as coke, organic milk, and other liquids are contained in PLA-made bottles.

In order to reduce landfilling, consumers are also very interested in eco-friendly drink bottles. Regrettably, certain products of that kind are not without limits. For industrial composting, PLA must be broken down. Furthermore, the polylactic acid bottle's heat resistance falls short of expectations. Plant-based PET or its mixed derivatives could prove more suitable for this kind of packaging due to certain factors.

Corn is one of the sources of the lactic acid that is utilized to produce recyclable bottles. Like polyglycolic acid or polymandelic acid, poly (lactic acid) (PLA) is a biodegradable and compostable member of the aliphatic polyester family. It is commonly generated from hydroxyl acids. In order to make products for the biocompatible/bioabsorbable medical device industry as well as bottles and other food item packaging, PLA, a thermoplastic polymer with a high modulus and strength, may be made from yearly renewable resources.

Lactide is the monomer of polylactic acid, derived entirely from naturally replenishable resources. For the most part, it is made utilizing field corn's plant sugars as the raw material. Because monomeric polylactic acid accounts for less than 0.05% of the world's yearly maize output, it has negligible to no effect on local or global food chains. Corn is not necessary; all it requires is a source of sugar. Sugar cane, sugar beet, tapioca, and wheat are completely renewable resources that are readily available and reasonably priced as substitute sugar sources for lactic acids. Sugar cane, sugar beet, tapioca, and wheat are completely renewable resources that can be easily obtained and affordably priced to replace sugar in the manufacturing of lactic acid. Then, lactic acid is prepared by fermenting a broth of sugar and starch with various lactic acid bacteria (LAB). LAB is almost exclusively found in the dextrorotatory (D or L (+)) form, which is used to produce PLA through three different methods: (i) ring opening polymerization; (ii) solid state polymerization; and (iii) condensation of lactic acid.

PLA-based packaging has gained popularity lately because it is mechanically resistant and has good stiffness. To increase PLA's durability, polyesters similar as polybutylene adipate-co-terephthalate are occasionally utilized. PLA-based polymeric materials have better mechanical strength for hot liquid filing over PLA's glass transition temperature. It is a thermoplastic polymer made entirely of biobased materials that has strong mechanical qualities and composts. In light of this, it is among the most significant materials. Depending on the amount of cellulose acetate's acetylation degree and the composition of the blend, PLA can offer extremely promising stiffness and full biodegradability when combined with it. When compared to raw PLA, the PLA-reinforced plasticized cellulose acetate composite exhibits a high Young's modulus and high toughness. These characteristics are all significant in the packaging sector.

Plasticizer can be used to increase PLA's toughness and ductility for flexible packaging. At room temperature, glass transition temperature (T_g)

of PLA, yield stress reduction, and increased elongation at break are noted upon the addition of plasticizers. However, the increased elasticity of sheets and films depends on these characteristics. Triacetin, oligoethers, oligo lactic esters, and oligo adipic esters for PLA and its blends are a few potent plasticizers. Acetyl tributyl citrate is another. For both rigid and flexible packaging—which is typically used for liquid and pasty cosmetics—good barrier qualities of PLA must be very high. The PLA cosmetic bottles contain several sorts of inorganic additives and clays that might alter this feature. It's also crucial that the PLA-based tubes have the right thickness for cosmetic

PLA is a polymer that many brands utilize to make bottles and containers for storing liquids. ConAgra Foods and Starbucks are collaborating to make their coffee containers recyclable. Conagra Foods is unique in that they generate shrink film packaging materials (such as the interior lining of Starbucks coffee containers for enhanced thermal stability) from post-industrial recycled polylactic acid. They have been able to transform 260,000 pounds of resin from non-renewable sources (like PVC and PET) to PLA by working with its suppliers. The world's first shampoo bottle made of organic polylactic acid plastic was created by the packaging company Capardoni and the cosmetics business Hair O'right. The PLA shampoo bottle, that Hair O'right says is the first in the world, is aligning the company's packaging with its environmental objectives. Hair O'right intends to replace the HDPE bottles made of petrochemicals with PLA for all of its upcoming goods.

3.2.2 Cellulose

When it comes to packaging materials supplied sustainably, liquid items—like water, syrup, curry, ketchup, oil, sanitizer, and soap solution—present much more issues than solid products because of their wettability. In the past, glass bottles or ceramic pots were used to package liquid goods including milk, water, and liquid beverages. The weight of glass increases during transportation, adding to the overall cost of the process. Additionally, glass's brittleness hindered them, and polymers quickly replaced glass packaging as plastic became widely available because of its customized qualities and inexpensive manufacturing. In addition, it was shown that compared to glass bottles, plastic bottles made of PET and high-density polyethylene (HDPE) have a smaller carbon footprint.

First-generation liquid packaging materials were glass and ceramic; second-generation materials were petroleum-based plastic; and currently, third-generation materials are required. In order to solve the drawbacks of the earlier materials with regard to ecological pollution and hence sustainably, this third-generation material should be bio-sourced and biodegradable.

Growing customer demands for more biobased products has resulted in advancements in the paper sector over time. These days, it's common knowledge that paper is used to package goods. Due to the difficulty in degrading that plastic products present, these products are a good substitute for their plastic counterparts. Products made of paper packaging may occasionally be recycled to yield cellulose fibres that can be reused. When recycling is not an option, paper products can nevertheless be disposed of in a way that is less damaging to the environment than plastics. They provide an affordable, effective, and adaptable means of safeguarding and moving a wide range of goods. They can also be customized to meet the needs of the product and are lightweight.

Cellulose, which is derived from plant-based sources that can include both wood and non-wood sources, is the most prevalent naturally occurring biopolymer. Furthermore, microorganisms can be used to produce cellulose, which is a more environmentally friendly method of production that prevents deforestation. Product packaging currently uses cellulosic material found in paper, cardboard, and moulded pulp. The most popular examples are the flexible packaging materials Tetra Pak for liquids, moulded pulp for trays, and layered laminate with another polymer.

Due to cellulose's hygroscopic nature, weak moisture and gas barrier qualities, cellulose-based materials are problematic when used to package liquid items. Consequently, cellulose is not the only material employed in liquid packaging. Aluminium provides the necessary barrier qualities to the paper-based packaging, while wax coating gives it hydrophobicity. It is also usual practice to use a heat seal polymer, such as polyethylene (PE), because paper is not actually a sealable substance. The paper structure is additionally protected from moisture by a layer of PE.

In contrast to milk or soap solutions, which are alkaline, ketchup and the majority of beverages are acidic. It follows that for various liquid items, the packing material needs to be resistant to acid and/or alkali. In milk's case, UV resistance is crucial to preserving the milk's nutritional value during storage in addition to its alkaline content. It is necessary to use packaging that is resistant to oil and does not change colour when it comes into touch with oil products, such as vegetable oil and motor oil, or stews that contain vegetable oil. Furthermore, because they contain volatile alcohols, hand sanitizers and alcoholic beverages may leak out of a packaging if it is not well-sealed and resilient to alcohols.

Polyethylene (PE), polypropylene (PP), or polyethylene terephthalate (PET) are the plastics used in the bottles; these materials are created using the blow moulding process. Because of their small weight, inexpensive production costs, and improved shelf visibility, these polymers took the place of bulky glass jars and tin cans. These sturdy, rigid plastic packaging are capable of supporting large loads. The most common types of rigid packaging include canisters, bottles, cans, clamshells, and cartons. But subsequently, as flexible packaging became more widely available, rigid packaging gradually began to disappear from the stores. Due to benefits such being lightweight, forming a customized barrier, requiring less raw materials for the same volume, requiring less energy than rigid packaging, being aesthetically pleasing, and many more, these flexible packages have supplanted rigid packaging. Usually composed of three layers, this flexible pouch packaging is printed on the first layer, which is the outermost layer. Aluminium is typically used to metalize the middle layer in order to provide an oxygen and moisture barrier. Finally, a sealant layer that comes into direct touch with food makes up the third and innermost layer. Some of these packaging involving the use of paper are discussed below.

3.2.3 Paper Sachets and Pouches

Multilayer laminate is used to seal flexible pouches and sachets, and heat is used to do this. The paper printing layer and the polymeric film sealing layer are the two common components of paper sachets. Aluminium foil, barrier polymer, or barrier coating can be used as a sandwich layer in this laminate. Only juices (tetrapak) can be packaged in paper pouches with aluminium foil acting as a barrier layer. Though many of them are designed for the packaging of solid objects, paper sachets and pouches have seen significant advancements. Lamitubes are the only liquid product format that may be used for toothpaste, makeup, and face creams. Lamitubes were invented by a few packaging companies, like Uflex, BillerudKorsnäs and Toppan, for use in various emulsified items, including toothpaste, hand wash, and cosmetics. Several packaging types, such as stand-up pouches, are created by laminating the paper base with a suitable polymer film and heat sealing it to create a pouch shape, in addition to these paper lamitubes. Nevertheless, these stand-up paper pouches produce waste since they contain an interior PE or PP sealant layer that prevents decomposition.

3.2.4 Paper Bottles

Polypropylene and polyethylene terephthalate are used to make water bottles. Since the majority of them are only intended for one-time usage, a significant quantity of waste is produced globally. In addition to water, bottles hold a variety of liquids, including milk, shampoo, soap in liquid form, fizzy, alcoholic, and non-alcoholic beverages, syrups (retail or pharmaceutical), and much more. Many sectors have started working on paper bottles since a change in packaging is required to transition to a more sustainable strategy for minimizing plastic waste. Paper bottles were first conceptualized as far back as 1908. It was addressed how the paper bottle could lower the cost of milk delivery and stop the spread of infectious illnesses.

Later, the tetra pack paper beverage carton was widely used, with polyethylene serving as the inner liner and aluminium serving as the barrier. Paper bottles are becoming more and more popular as a replacement for traditional bottles. Recycled pulp is used to make the outer layer of every paper bottle project undertaken by businesses globally. Each bottle has a separate inner layer made of partially conventional polymers, partially bio-sourced polymers, partially biobased polymers, and in certain cases, partially proprietary coatings that are not published. Although these inner liners are food grade, they don't have any barriers or particular uses, such as carbon dioxide barrier, alcohol resistance, or hot fill. Here are a few examples of bottle packaging that has made use of paper.

3.2.4.1 Eco Bottle

The paper-based exterior shell of the Eco.Bottle (USA) water bottle is composed of recycled cardboard or newspaper, which may be recycled once more. Standard recycled plastic is used to make the inside shell. These outer and inner shells may be readily separated and transferred to different recyclable streams because neither layer is laminated. A unique procedure covers an interior polymer liner with an exterior shell that is created in two sections. Because pulp is used in the bottle, traditional polymer is used with less thickness, resulting in as much as 60% less polymer material being used.

3.2.4.2 Paper Water Bottle by Eco1green

Using a proprietary technology, Eco1green (USA) created a sustainable bottle composed of 35% PET and 65% pulp. This layer of PET is called Enso Restore PET, an addition produced by Enso Plastics that speeds up the natural breakdown of PET when added to regular PET. It takes one to 15 years for the degradation to happen, and it only happens when natural microbial activity is landfilled. Eco1Green claims that its bottle is 65% compostable (pulp), 98% biodegradable in landfills, and 100% recyclable. Additionally, their goal is to have 100% compostable bottles.

3.2.4.3 Frugal Bottles

While the inner barrier liner of Frugal Bottle is made of metalized laminate, the exterior shell is still made of paper. This laminate is made up of two layers: PE for food contact and sealing, and metalized PET for a barrier. For simplicity of recycling, the shell's design allows both layers to be split simply by applying pressure at a specific spot. It is said to be five times lighter than a glass bottle, use 77% less plastic than a standard bottle, and have a carbon footprint that is six times smaller than a glass bottle.

3.2.4.4 Puplex

Pilot Lite and Diageo founded the bottling company Pulpex (England). They started producing alcohol bottles (Johny Walker) with an inside proprietary coating made of pulp. As of right now, the coating holds ordinary liquid and is compatible with bottles that accept hot filling and carbonated beverages. Additionally, they collaborated with Pepsico, GSK Healthcare, Castrol, Uniliver, Stora Enso, and Solenis to create, expand, and enhance the paper container, serving as a barrier for a range of uses.

3.2.4.5 Kagzi

Kagzi (India) is a TableBandi LLC company that produces paper bottles with a proprietary material coating and bottles that are purportedly plastic-free. The pulp is formed into the two sides of a bottle using recycled paper and paperboard. After that, the unpainted bottle is spray coated and assembled to create a bottle. Within months, the inner layer is believed to dissolve into the soil and be 100% compostable [41].

When it comes to cutting carbon emissions and improving environmental performance, paper bottles have a lot going for them over other materials. For drinks and other liquid items like oils, there is a sustainable packaging solution that printers, packaging companies, and co-packers can benefit from. The market for paper bottles will breathe fresh life into the sector by creating new supply chains for packaging, print shops, and providers of recycled paper. The industry's dedication to lowering emissions and consumer desire for more environmentally friendly products might spark a revolution in paper bottles!

3.2.5 Polyethylene Furanoate (PEF)

Made entirely of plant-derived sugars, PEF is a plastic polymer. Cultivated crops including wheat, corn, and sugar beet can yield the sugars (fructose) needed to create furandicarboxylic acid (FDCA). The foundation of PEF is FDCA, a highly effective, plant-based, totally recyclable plastic. A 100% plant-based PEF polymer is created by polymerizing FDCA with mono-ethylene glycol (MEG) derived from plants. When technology reaches its peak, PEF can also be made from cellulose, which is widely available in non-edible biomass sources like forestry and agricultural waste streams. European wheat starch is used in the current technique. The other essential component is commercially accessible biobased MEG.

Compared to the petroleum-based polymers that are now commonly utilized, PEF offers improved mechanical, thermal, and barrier qualities. Produced at scale, PEF's barrier qualities—about 10 times better for O_2, 15 times better for CO_2, and 2.5 times better for water than PET—represent a revolutionary possibility when compared to conventional packaging options in terms of performance, cost, and sustainability. Packaged goods have an extended shelf life due to their enhanced barrier qualities. Furthermore, PEF has greater mechanical strength, allowing for the production of thinner PEF packaging with less resources needed. PEF is more heat-resistant than other materials and may be produced at lower temperatures due to its excellent thermal characteristics. PEF provides increased shape possibilities and improved mechanical rigidity.

In order to adequately satisfy the demands of the modern world, PEF's end-of-life and circularity are just as crucial as its plant-based source, performance, and cost. PEF can replace glass, aluminium cans, and multilayer bottles when utilized as a single layer in bottles for juices, soft drinks, and beer. PEF has shown to be functional with the current recycling and sorting facilities. PET recycling machinery can also be used to recycle PEF both chemically and mechanically in a same manner. PEF is a better option than PET. Among the notable PEF attributes are:

  i. Biodegradability: When composted industrially, PEF breaks down naturally. It decomposes far more quickly than PET and other polymers.
 ii. Gas barrier qualities: PET is inferior to PEF in this regard. PEF is therefore appropriate for packaging fizzy beverages. Gas escape prevention is crucial.
iii. Mechanical properties: PEF exhibits good mechanical qualities and an excellent oxygen barrier. Its stiffness is comparable to PET, but its tensile strength is greater. This makes thinner PEF bottles possible, conserving material.
 iv. Thermal characteristics: PEF can tolerate hotter temperatures. Compared to PET, it has a greater melting point, modulus, and glass transition temperature.
  v. Renewability: Plant sugars that are renewable are used by PEF. Fossil fuels are used by PET. PEF benefits from sustainability in this way. When biobased MEG is used, FDCA-based PEF is 100% biobased.

vi. Recyclability: PEF has the potential to be recycled, however infrastructure needs to be developed. Integrating PEF recycling with the current PET recycling program is the aim.

The residual plant matter from pressing sugarcane is the first step in the process. PEF is the polymeric polymer that results from a series of reaction stages that include the addition of some collected CO_2 and some ethylene glycol made from maize plants. It functions similarly to PET plastic, which is represented by the number one recycling sign in water and soda bottles. Around one-third fewer greenhouse gases are released during the PEF process than during PET production (Figure 3.1).

This depends on the assumption that natural gas, rather than renewable energy sources, provides the heat and electricity needed for production. However, some of the CO_2 that has been absorbed throughout the process is consumed, balancing some emissions. Remarkably, emissions from other suggested PEF production techniques are actually lower than that. But the researchers intended to avoid using leftover plant material, since such methods relied on using food sugars instead. Figure 3.1 depicts the production of PEF bottles [42].

Since PEF is a fairly recent material, it is not yet offered for sale. PET is a proven product with over 40 years of market experience, produced on a massive scale using a highly established technology that operates at nearly maximum efficiency. Significant technological, economic, and environmental improvements spanning the whole value chain are anticipated as a result of the PEF market's commercialization and expansion.

Scientists are looking into using a variety of biopolymers, including chitin, chitosan, zein, collagen, and hyaluronic acid, to replace the plastic bottles that are now in use in the manufacturing of bottles.

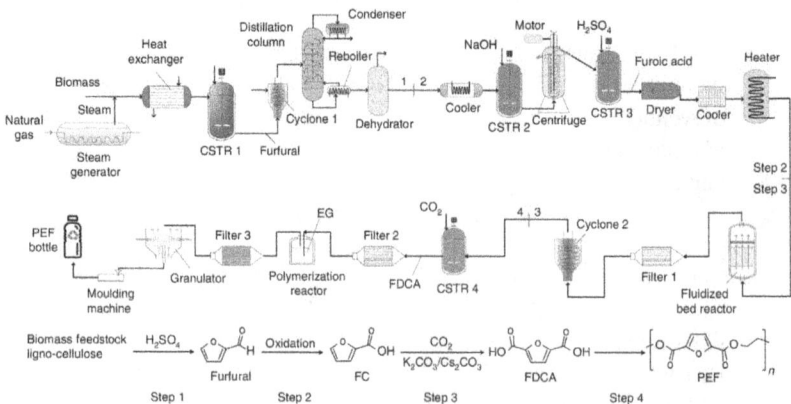

FIGURE 3.1 Preparation of PEF bottles

CHAPTER 3.3: ADVANTAGES OF BIOPOLYMERS APPLICATIONS IN BOTTLE PACKAGING

The potential of biopolymers to replace traditional plastics in a range of applications, such as the creation of reusable bottles, has drawn attention. There are several benefits to using biopolymers in packaging, particularly when considering environmental effect and sustainability. The principal benefits are as follows:

a. **Material Composition**: Typically, biopolymers used to make bottles come from renewable resources like cellulose, sugarcane, or maize starch. Biopolymers like polylactic acid (PLA) and polyhydroxyalkanoates (PHA) are frequently employed for this purpose.

b. **Biodegradability and Compostability**: In contrast to conventional plastics made from fossil fuels (like PET), biopolymers are frequently compostable and biodegradable in the right circumstances. This implies that they have the ability to decompose into non-toxic elements in particular settings, so mitigating their influence on the environment.

c. **Recyclability**: Biopolymers can be made to be recyclable, which enables them to be gathered, processed, and utilized again to make new goods. By continuously reusing things instead of discarding them after a single use, this encourages the development of a circular economy.

d. **Mechanical Properties**: Biopolymers are suited for a range of packaging applications, including bottles, since they may be developed to have qualities like strength, flexibility, and transparency that are comparable to those of conventional plastics.

e. **Environmental Benefits**: Utilizing biopolymers lowers greenhouse gas emissions related to the production of plastic and lessens reliance on finite fossil resources. They also help lessen the amount of plastic garbage that ends up in the environment, particularly if they are made to be recyclable or biodegradable.

f. **The possibility of consuming less energy**: Some biopolymers contribute to overall energy savings by requiring less energy for production than conventional plastics.

g. **Non-toxic**: Biopolymers usually don't produce toxic compounds as they break down, which is good for the environment and people's health.

h. **Improved Brand Image**: Employing biopolymers can help a business project a more sustainable and ecologically conscious image, which will appeal to customers who are becoming more conscious of environmental effects.

i. **Functional advantages**: Biopolymers can be developed to possess particular functional features, including as strength, flexibility, and gas and moisture barrier qualities, making them appropriate for a range of packaging applications.

j. **Regulatory Aid**: Businesses are encouraged to employ biopolymers by regulations that restrict the use of regular plastics and encourage the use of biodegradable or compostable materials.

k. **Waste Reduction**: By promoting composting or biodegradation, biopolymers might help reduce waste generation and possibly complete the circle in the framework of the circular economy.

l. **Innovation Potential**: As biopolymer technology continues to be researched and developed, new material qualities could be improved, increasing the range of applications for the materials outside of packaging.

m. **Food compatibility**: Natural biopolymers barely interact with food and are not harmful when consumed.

All things considered, employing biopolymers in packaging has a positive impact on the environment, advances sustainable development objectives, and satisfies customer demand for green products. Even if there are still issues with cost and performance inconsistency, continued development and wider use should help to make biopolymers more viable and competitive in the packaging industry.

CHAPTER 3.4: LIMITATIONS OF BIOPOLYMER USAGE

There are several benefits to using biopolymers in the production of bottles, but there are drawbacks and difficulties as well:

a. **Cost**: When compared to conventional plastics derived from petroleum, such as PET (polyethylene terephthalate), biopolymers are typically more expensive to produce. This cost difference results from various factors, including production procedures, economies of scale, and the cost of raw materials (such as agricultural feedstocks).

b. **Performance Variability**: When it comes to mechanical and barrier qualities, biopolymers can differ from conventional plastics. This variation is contingent upon variables like the particular kind of biopolymer, techniques of processing, and ambient circumstances. It can be difficult to achieve consistent performance across many applications.

c. **Processing Constraints**: When it comes to processing, biopolymers frequently need different methods than traditional plastics. This entails using diverse additives to attain the desired qualities, executing processes at slower speeds, and lower temperatures during processing to prevent degradation. The process of modifying the current production system to incorporate biopolymers can be expensive and intricate.

d. **Durability and Shelf Life**: In comparison to traditional plastics, some biopolymers could not be as strong or long-lasting, particularly when it comes to retaining their mechanical properties and barrier qualities over time or in adverse environmental circumstances. This restriction may limit the use of biopolymers in some applications, such long-term packaging.

e. **Compatibility with Recycling Infrastructure**: Although biopolymers can be made to be recyclable, there may be limitations to how well they work with the current recycling system, which is mostly made for conventional plastics. Logistically challenging tasks include separating biopolymers from conventional plastics and making sure that recycling procedures are effective and free of contaminants.

f. **Land Use and Competition with Food Production**: A lot of biopolymers are made from feedstocks used in agriculture, like corn and sugarcane. Land use, rivalry with food production, and possible effects on food availability and costs are raised by this, particularly if large-scale biopolymer manufacturing increases.

g. **Environmental Impact Considerations**: Although biopolymers have the potential to be environmentally beneficial, their total impact can differ based on factors like where raw materials are sourced, how they are produced, and what happens to them after their useful life—such as the availability of composting facilities. To accurately compare the environmental impact of biopolymers to traditional plastics, life cycle studies are essential.

In order to overcome these constraints, continued research and development is needed to enhance the characteristics of biopolymers, increase production efficiency, increase recycling capacity, and guarantee sustainable sourcing methods. Biopolymers continue to be a viable substitute for conventional plastics in applications such as the production of bottles, in spite of these obstacles. This helps to lessen the waste produced by plastic and lessen dependency on fossil fuels.

Regulatory Framework and Policies

4

CHAPTER 4.1: GLOBAL REGULATIONS ON PLASTIC BOTTLE PRODUCTION AND DISPOSAL

Plastic bottle production and disposal are globally regulated to mitigate environmental impacts and promote sustainability across the product lifecycle. As ubiquitous containers for beverages and other liquids, plastic bottles present significant challenges due to their persistence in the environment and potential harm to wildlife and ecosystems. International regulations and policies address various aspects of plastic bottle production, including material standards, recycling requirements, and waste management practices. These regulations aim to minimize plastic pollution, encourage resource efficiency, and promote the adoption of circular economy principles. Compliance with these regulations is crucial for ensuring the responsible manufacture, use, and disposal of plastic bottles worldwide, contributing to efforts aimed at achieving environmental sustainability and reducing the global carbon footprint.

4.1.1 India

4.1.1.1 BIS Regulations on Plastic Bottles

The BIS is the National Standard Body of India, created by the BIS Act of 2016 to promote the orderly growth of the activities related to standardization, product certification, and marking, as well as any related or incidental problems. Through standardization, certification, and testing, BIS has been able to control the proliferation of varieties, promote exports and imports as alternatives, minimize health hazards to consumers, and provide safe, reliable, high-quality goods, among other accountability and tangibility benefits that boost the national economy.

BIS created two Indian Standards: IS 14543:2016, which is the Standard for Bottled Water for Drinking (other than Packaged Natural Mineral Water), and IS 13428:2005, which is the Specification for Packaged Natural Mineral Water. As per the BIS Conformity Assessment Regulations, 2018, Scheme I of Schedule II, the organization certifies products. To date, it has authorized more than 6,200 licenses for IS 14543 and more than 30 licenses for IS 13,428 to different manufacturing units, including both foreign and domestic producers. Bottled drinking water (other than natural mineral water) as per IS 14543:2016 and Packaged Natural Mineral Water as per IS 13428:2005 are subject to required certification by BIS under the Food Safety and Standards (Prohibition and Restriction on Sales) Regulation, 2011.

Therefore, it is illegal for anybody to produce, market, or display bottled drinking water and natural mineral water for sale unless they are certified by the Bureau of Indian Standards.

In order to ensure a safe, hygienic, and nutritious product, IS 14543-1998 (Specification for Packaged Drinking Water) specifies the hygienic practices to be followed with regard to water collection, treatment, bottling, storage, packaging, transport, distribution, and sale for direct consumption.

The labelling of bottles completes the process of packing drinking water. The Ministry of Health and Family Welfare, along with the Department of Health, have issued a notification advising disposable mineral water or packaged drinking water labels should read "Crush the bottle after use." Additional labelling specifications should follow PFA Regulations and Packaged Commodities Regulations as stated in IS 14543-1998 (Specification for Packaged Drinking Water). The table below includes a list of various BIS numbers along with their titles [43] (Table 4.1).

4.1.1.2 Disposal of Plastic Bottles as per CPCB

In India, the Central Pollution Control Board (CPCB) is essential to monitoring and controlling plastic waste management procedures all over the nation. The CPCB, the highest regulatory agency within the Ministry of Environment, Forests, and Climate Change (MoEFCC), is in charge of developing rules and carrying out programs to deal with the problems that plastic waste poses to the environment. Its mandate includes monitoring and enforcing compliance with Plastic Waste Management Rules, which set standards

TABLE 4.1 BIS titles on plastic packaging in India

IS NUMBER	IS TITLE
10142:1999	Specification for the safe use of polystyrene (crystal and high impact) in contact with food, medication, and drinking water (*First Revision*)
10146:1982	Guidelines for the safe use of polyethylene in contact with food, medication, and drinking water
10148:2023	Positive list of ingredients for polyvinyl chloride (PVC) and its copolymers that come into contact with prescription drugs, food, and water (*First Revision*)
10149:1982	Positive list of styrene polymer compounds in interaction with food, medicine, and drinking water
10151:2019	The specification for polyvinyl chloride (PVC) and its copolymers for safe use in contact with food, medicine, and drinking water (*First Revision*)
10910:1984	Guidelines pertaining to the safe handling of polypropylene and its copolymers in relation to food, medicine, and drinking water
11434:2023	The specifications for isomer resins that can be used safely in contact with food, medication, and drinking water (*First Revision*)
11435:2024	Positive list of ionomer resin ingredients that are safe to use when in contact with food, medicine, and water (*First Revision*)
14971:2001	Specification for the safe use of polycarbonate resins in contact with food, medication, and drinking water
15410:2003	Specifications for containers used to package drinking water and natural mineral water
16621:2017	A positive list of the ingredients in polyethylene and polypropylene that come into contact with food, medicine, and drinking water
14625:2015	Feeding bottle made of plastic (*First Revision*)
12252:1987	For safe usage in contact with food and drinking water, use PET (polyethylene terephthalate) or PBT (polybutadiene terephthalate)
9833:1981	Enumeration of colours and pigments that should not be used in plastic that comes into contact with food or beverages. In essence, the standards outline specifications for the fundamental resin, colours and pigments, emulsifying agents, catalysts, residual monomers, antioxidants, additional additives, and overall migration.
14543:2016	Specification for Packaged Drinking Water (Except Bottled Natural Mineral Water) (*Second Revision*)

for the collection, segregation, recycling, and disposal of plastic waste. The CPCB collaborates closely with state pollution control boards (SPCBs), municipal authorities, and other stakeholders to ensure effective implementation of these regulations. Through initiatives aimed at promoting awareness, fostering innovation in recycling technologies, and advocating for sustainable practices, the CPCB endeavours to mitigate the environmental impact of plastic waste while advancing India's goals towards a cleaner and more sustainable future.

The former Plastic trash (Management & Handling) Rules, 2011 have been amended by the Government of India, which on March 18, 2016, notified the Plastic Waste Management (PWM) Rules, 2016, for the efficient management of plastic trash. All waste generators, local bodies, gram panchayats, manufacturers, importers, producers, and brand owners in India would be subject to these regulations (Table 4.2).

Technology is encouraged to be used for the disposal of plastic waste according to PWM Rules, 2016 clause 5(b). The following is a discussion of the main technologies for disposing of plastic garbage.

4.1.1.2.1 Using Plastic Waste in the Construction of Roads

With the exception of chlorinated and brominated plastic trash, mixed MSW is collected and separated into different categories. The separated plastic garbage is kept in storage and needs to be brought to the designated work site so that it can dry. After being dried and chopped into 2–4 mm pieces, the plastic waste is combined with hot stone aggregate and mixed thoroughly. Additionally, heated bitumen—which is used for compaction and laying—is combined with the coated aggregate. Currently, a number of States and Union Territories, including Tamil Nadu, Himachal Pradesh, Nagaland, West Bengal, Pondicherry, etc., have built highways utilizing plastic trash and asphalt.

TABLE 4.2 Responsibilities of CBCP as per PWM

RULE NUMBER (AS PER PWM)	DESCRIPTION
5(c)	The disposal and processing of thermoset plastic waste must adhere to the instructions periodically released by the CPCB.
6(2)(d)	Local Bodies are responsible for processing and disposing of the non-recyclable portion of plastic trash in compliance with the CPCB's standards.
17(d)	Every year by August 31st, the CPCB will have prepared a comprehensive Annual Report on the use and management of plastic trash and would transmit it to the Central Government with its recommendations.

4.1.1.2.2 Co-processing Plastic Waste with Cement Kiln Processing
Utilizing waste materials as alternative fuels and raw materials (AFR) in industrial processes to extract material and energy from them is known as co-processing. The cement kiln's high temperature and extended residence duration allow for the efficient disposal of all waste kinds without producing any hazardous pollutants. According to the Basal Convention, a cement kiln's co-processing technology allows for the environmentally sound and safe disposal of a variety of pollutants, including hazardous wastes. Plastic trash is utilized as an Alternative Fuel and Raw Material (AFR) in cement factories, where it is heated to a temperature of between 1,400°C and 1,500°C. Energy is recovered during the process as plastic trash burns and the inorganic content is fixed with clinker. Cement kilns need to be fed plastic trash using an automated feeding system. Several states, including Gujarat, Tamil Nadu, Karnataka, Chhattisgarh, Himachal Pradesh, Madhya Pradesh, Odisha, and others, have cement factories with the capacity to co-process waste, and these states effectively employ this technology.

4.1.1.2.3 Recycling Plastic Waste to Produce Fuel
Oil: Refused-Derived Fuel (RDF)
The purpose of collecting and sorting plastic waste is to turn it into fuel oil (RDF). After being separated, the plastic waste is put into a multi fractionalization process, where the undesirable material is discarded in order to facilitate easier handling and processing. Subsequently, the depolymerization system within the vessel receives the separated plastic trash (which solely consists of HDPE, LDPE, PP, and multi-layer packaging save PVC). Polymers are intended to be handled by the Catalytic Gasolysis in-vessel. The choice of raw material determines the catalyst to be employed. Under airless conditions, the reactor runs at a high temperature. Gasolysis of the polymers to tiny chain hydrocarbon linkage occurs at high temperature. As crude oil is collected, the generated vapours condense in the condensers. Three different types of condensation occur: the first produces fuel oil (FO), the second produces light diesel oil (LDO), and the third produces the highest-grade diesel oil. Depending on the quality of the plastics and contaminations entered, the overall percentage of this is often between 40% and 50% of the input. After that, the non-condensable residues are sent through a scrubber to remove gases like gasoline and chlorine, among others. This Gas-Fuel is utilized for heating in the process. A few municipalities, like NDMC (New Delhi) and Vadodara (Gujarat), use this technique.

4.1.1.2.4 Using Plasma Pyrolysis Technology (PPT)
to Dispose of Waste Plastic
The process of breaking down organic and inorganic materials into gases and non-leachable solid wastes in an oxygen-starved atmosphere is known as

plasma pyrolysis technology. Large fractions of electrons, ions, and excited molecules are used in plasma pyrolysis along with high intensity radiation to break down compounds. In this procedure, molecular bonds are broken using plasmas, the fourth state of matter. PPT can be used to dispose of a variety of plastic waste materials, including polyethylene bags, filthy plastic, metalized plastic, multi-layer plastic, and PVC plastic.

The initial step in plasma pyrolysis is feeding the waste plastic through a feeder into the main chamber at 850 degrees Celsius. The waste material breaks down into higher hydrocarbons, methane, hydrogen, and carbon monoxide, among other things. The debris from plastics and the pyrolysis gases are drained into the secondary chamber by an induced draft fan, where they burn in the presence of more air. High voltage spark is used to ignite the combustible gasses. 1,050°C is the temperature that is kept in the secondary chamber. Safe carbon dioxide and water are produced from the combustion of the hydrocarbon, CO, and hydrogen. In the case of chlorinated waste, the process parameters are kept so that the creation of hazardous dioxins and furan molecules is completely eliminated. Few municipalities and hospitals utilize this procedure, but it might be helpful in remote locations such as hill stations, tourist destinations, pilgrimage sites, and coastlines.

4.1.2 USA

4.1.2.1 US FDA Regulations of Plastic Bottle Packaging

The United States Food and Drug Administration (FDA) is a federal agency within the Department of Health and Human Services (HHS). The FDA is responsible for protecting public health by regulating and supervising a wide range of products. Drinking water safety is under the jurisdiction of both the Environmental Protection Agency (EPA) and the FDA. While the FDA controls bottled drinking water, the EPA governs tap water used for public use. As per FDA, a consumable's contact material's overall regulatory status is determined by the regulations governing each of the constituent substances that make up the article. Because of its intended use in the alimentary contact material, any individual substance that is reasonably predicted to migrate to foodstuff will be covered by one of the following:

- a rule found in the Federal Regulations Title 21 Code
- fulfilling the requirements for Generally Recognized As Safe (GRAS) status, which may include fulfilling a GRAS notice or regulation.

- an earlier letter of reprimand
- a request for a Threshold of Regulation (TOR) exemption
- a Food Contact Substance Notification (FCN) that works

The United States Food and Drug Administration (FDA) has issued important regulations that are contained in Title 21 of the Code of Federal Regulations (CFR). It focuses on food and medication in particular, encompassing a wide range of subjects necessary to guarantee the efficacy, safety, and quality of goods pertaining to public health. Aspects of FDA-regulated items such as food additives, packaging materials, medications, medical devices, cosmetics, nutritional supplements, and more are covered under the several sections of Title 21 CFR. It provides producers, distributors, healthcare practitioners, and consumers with a thorough handbook that outlines the standards, rules, and processes needed to adhere to FDA laws and guarantee the safety and health of the general population.

The US FDA regulates the bottling and processing of bottled drinking water under Title 21 CFR Part 129. In order to guarantee the safety and quality of packaged water during its manufacturing, storage, and delivery, it sets forth extensive specifications. This covers criteria for good manufacturing procedures (GMP), labelling, sanitation practices, processing techniques, and source water quality standards. By guaranteeing that bottled drinking water is free of impurities and satisfies strict FDA regulations prior to reaching consumers, Part 129 seeks to protect the public's health. Manufacturers and distributors must abide by these rules in order to maintain the reliability and security of bottled water products in the marketplace.

In the US FDA rules, Title 21 CFR Part 170 describes the broad guidelines and processes for evaluating the safety of consumable additives. The framework it offers allows for the assessment of the safety of materials that are purposefully introduced to food, such as those found in packaging. In order to comply with Part 170, firms must provide scientific evidence proving food additives do not endanger consumers when used as intended. For compounds deemed generally recognized as safe (GRAS), the rule addresses the filing of notifications and petitions for food additives. In order to contribute to the overall protection of public health, Part 170 compliance guarantees that food additives, including those in packaging materials, satisfy FDA requirements for safety.

The US FDA's Title 21 CFR Part 175 regulation covers indirect food additives; it focuses on adhesives, coatings, and other materials used in food packaging. In order to guarantee that these materials are safe for their intended use and do not contaminate food—including water in packaging—with hazardous compounds, this regulation lays out standards. The use of these additives is governed by Part 175's regulations, which also set maximum permissible amounts and composition restrictions. Manufacturers must keep documentation attesting to their adherence to FDA regulations and perform safety assessments.

21 CFR Title The US FDA's regulations address the safety of paper and paperboard components used in food packaging. It lays out the guidelines and circumstances that must be followed in order to use these materials properly and prevent the contamination of food—including water in packaging—with hazardous compounds. Paper and paperboard must adhere to FDA regulations for materials that come into contact with food, and Part 176 specifies what coatings, treatments, and additives are acceptable. In order to prove compliance, manufacturers are required to undertake safety evaluations and keep records. By reducing the possibility of contamination from paper-based food packaging materials, such as bottled water, this law is essential to protecting the public's health.

21 CFR Title of the US FDA's Part 177 standards control the safety of plastics and polymers used as indirect food additives in food packaging materials, such as water containers. It outlines specifications to guarantee that these materials are secure and do not carry a danger of contaminating food. Conditions governing the kinds of polymers and plastics that can be utilized, along with approved processing techniques and additives, are outlined in Part 177.

The US FDA regulates the use of materials as indirect food additives under Title 21 CFR Part 178. This includes the use of adjuvants, manufacturing aids, and sanitizers in food packaging materials. In order to guarantee that these drugs are safe for their intended use and do not endanger consumers' health, this rule lays out standards. Part 178 specifies permitted materials, how they may be used, and sets safety guidelines including maximum amounts and proper manufacturing techniques. In order to prove compliance with FDA standards, manufacturers must maintain records, follow good manufacturing practices (GMP), and undertake safety reviews (Table 4.3).

4.1.2.2 Disposal of Plastic Bottles as per Environmental Protection Agency (EPA)

The Environmental Protection Agency (EPA) in the USA does not have specific federal regulations solely dedicated to the disposal of plastic bottles. However, the EPA plays a significant role in guiding and supporting states and local governments in managing solid waste, including plastic bottles, through various environmental programs and initiatives. The EPA promotes recycling as the preferable way to manage plastic bottles and other recyclables and pushes the "reduce, reuse, recycle" principles to decrease trash output. The Resource Conservation and Recovery Act (RCRA) gives the EPA

TABLE 4.3 FDA titles on plastic packaging of consumable items

FDA TITLE 21 CFR PART	DESCRIPTION
129	Governs the processing and bottling of bottled drinking water, ensuring its safety and quality from production to packaging.
170	Provides the general principles and procedures for determining the safety of food additives, including those used in packaging materials for water and other foods.
175	Regulates indirect food additives such as adhesives, coatings, and other substances that may come into contact with water and food packaging.
176	Addresses the safety requirements for paper and paperboard components used in food packaging, including those used for bottled water.
177	Deals with polymers and plastics used as indirect food additives in packaging materials, ensuring their safety for use with water and other foods.
178	Covers the use of adjuvants, production aids, and sanitizers in food packaging materials, ensuring they do not pose a health risk to consumers.

control over the management of both hazardous and non-hazardous solid waste. Despite the fact that plastic bottles alone are typically not regarded as hazardous waste, RCRA sets the standards for waste management to protect the environment and public health. In order to lessen the negative effects of plastic bottles on the environment, the EPA is in favour of programs that reduce litter, stop marine debris, and encourage sustainable materials management techniques [44].

4.1.3 European Union

4.1.3.1 EU Regulations of Plastic Bottle Packaging

Laws pertaining to food and packaging in the European Union (EU) are mainly created and supervised by a number of important organizations. Legislation proposals, policy implementation, and law enforcement within the EU fall under the purview of the European Commission (EC). Regulations pertaining to waste management, packaging materials, and food safety are developed and adopted with its help. Scientific guidance on matters pertaining to food safety is provided by the European Food Safety Authority (EFSA), an independent EU institution. It evaluates food chain risks, such as chemicals used in packing materials that could end up in food. The European Commission's proposed regulations are debated and adopted by the European Parliament and Council of the European Union as part of the legislative process. Through the regular legislative process, they are instrumental in drafting regulations pertaining to food and packaging. To top it all off, each EU member state has authorized authorities tasked with implementing EU laws within its borders, including keeping an eye on adherence to packaging and food safety guidelines. These organizations work together to create and implement comprehensive laws that provide the European Union's food products and packaging materials with sustainability, safety, and quality assurance.

Within the European Union (EU), various kinds of plastics and bottle packaging are regulated by a number of laws and regulations passed by the European Commission (EC). The goals of these laws are to protect consumer safety, advance sustainability, and reduce the environmental effect of packaging materials. The main EC legislation and rules pertaining to particular kinds of plastics and bottle packaging are as follows:

1. **Directive 94/62/EC on Packaging and Packaging Waste**: This regulation establishes the fundamental specifications for packaging that is sold within the European Union. With the goal of minimizing the environmental impact of packaging waste through reduction, reuse, and recycling, it includes rules on the types of polymers that can be used in package materials.

2. **Directive (EU) 2019/904 on Single-Use Plastics**: This regulation specifically targets plastic bottles and other single-use items made of plastic. It prohibits some single-use plastic products from being sold and establishes policies to cut back on usage, encourage substitutions, and guarantee recyclable materials.

3. **Regulation (EU) No 10/2011 on Plastic Materials and Articles Intended to Come into Contact with Food**: Food-contact materials, such as plastic bottles, are subject to regulations regarding the kinds of polymers and plastic additives that can be utilized. Through guaranteeing the security of food packaging materials, it seeks to safeguard consumer health.

4. **Directive 2002/72/EC relating to Plastic Materials and Articles Intended to Come into Contact with Food**: Specific guidelines for the composition and use of plastic materials in food contact applications, such as plastic bottles, were established by this directive, which is now superseded by Regulation (EU) No 10/2011.

5. **Regulation (EU) No 282/2008 on Recycled Plastic Materials and Articles Intended to Come into Contact with Food**: The types of recycled plastics that can be utilized in plastic bottles and other packaging are among the requirements for the use of recycled plastics in food-contact materials that are outlined in this regulation.

6. **Regulation (EU) 2018/852 on Packaging and Packaging Waste:** The recycling goals set forth by this legislation are high for packaging materials, especially plastic bottles. It mandates that actions be taken by EU member states to guarantee that recycling and recovery goals are fulfilled and that packaging waste is decreased.

4.1.3.2 Disposal of Plastic Bottles as per European Union

In order to promote recycling, minimize waste, and minimize environmental effect, the European Union (EU) has put in place a system of policies and standards that govern how plastic bottles are disposed of. Plastic bottle and other packaging material recycling programs must be established and carried out by EU member states. The Packaging and Packaging Waste Directive (94/62/EC) establishes goals for the recovery and recycling of packaging waste, with particular attention paid to plastics. A specific proportion of waste plastic packaging must be collected or recycled, as mandated by member states under the Deposit Return Schemes (DRS). Consumers that use DRS for plastic bottles in certain EU member states pay a deposit that is refunded upon the bottle's return for recycling. Reducing littering and raising recycling rates are two benefits of DRS. By 2030, all plastic packaging sold in the EU is expected to be recyclable or reused, according to the EU Plastic Strategy. Among its goals are the reduction of plastic waste, particularly plastic bottles, and the promotion of the use of recycled plastics in new goods. By 2030, all plastic packaging sold in the EU is expected to be recyclable or reused, according to the EU Plastic Strategy. Among its goals are the reduction of plastic waste, particularly plastic bottles, and the promotion of the use of recycled plastics in new goods. Overall, the EU framework aims to promote a circular economy approach where resources are used more sustainably, waste is minimized, and environmental impacts are reduced throughout the lifecycle of products, including plastic bottles. Each member state is responsible for implementing and enforcing these regulations within their national context, ensuring consistent standards across the EU [45].

CHAPTER 4.2: EFFICACY OF IMPLEMENTING EXTENDED PRODUCER RESPONSIBILITY (EPR)

A regulation strategy known as Extended Producer Responsibility (EPR) requires importers, manufacturers, and occasionally retailers to assume accountability for the environmental effects of their goods at every stage of their lifespan, from production to disposal. By transferring the conventional duty of waste management from government agencies to manufacturers, this idea encourages manufacturers to create products that are safer, easier to recycle, or reuse. When a product is no longer valuable to customers, producers are usually required to fund and manage programs for its collection, recycling, and disposal. This entails financial and operational obligations for producers. EPR seeks to lessen waste production, decrease environmental degradation, and promote the shift to a circular economy—one in which materials are reused and recycled—by encouraging sustainable product design and resource efficiency.

Organization for Economic Co-operation and Development, or OECD, is an international organization that serves as a forum for governments to work together to address global economic, social, and environmental challenges. Established in 1961 and headquartered in Paris, France, the 38 member nations of the OECD are primarily from North America and Europe. Although the European Union (EU) was the first region to enact EPR laws, they are now in place throughout the world, including in Canada, the US, Australia, Japan, and India. Despite being widespread, EPR law differs not only between nations but also between states, provinces, regions, and other geographical areas. This includes differences in schedules, reporting requirements, tariffs, and fee structures. With respect to Extended Producer Responsibility (EPR), the OECD is a major contributor by means of discussion facilitation, exchange of best practices, and advice on the administration and implementation of EPR programs. The OECD is a preeminent international forum that fosters dialogue on the creation and execution of EPR programs between governments, business stakeholders, and civil society. In order to assist nations in creating strong EPR frameworks that support sustainable patterns of production and consumption, it conducts research and provides policy recommendations. The following attributes of EPR policies are listed by the OECD:

- The upstream shift in accountability from municipalities to producers

- Offering incentives to manufacturers so they will take environmental effects into account while creating products

Although products of any kind may fall within the purview of EPR laws, batteries, packaging, and WEEE (waste electrical and electronic equipment) are the three main product categories. These groups have been ranked by legislators according to the quantity and toxicity of their waste streams.

The effectiveness of Extended Producer Responsibility (EPR) can be assessed by examining multiple crucial aspects that emphasize its effects on the environment, economy, and society. EPR, first and foremost, promotes manufacturers to create items that are easier to recycle and have a smaller environmental impact over the course of their lives. By encouraging the use of recycled materials in new products, this strategy serves to prevent resource depletion, reduce landfill trash, and raise recycling rates. Economically speaking, EPR encourages innovation in recycling technology and product design, which could result in cost savings through increased resource efficiency as well as the creation of jobs in the recycling sector. Socially, EPR promotes a culture of sustainability and accountability by increasing public understanding of waste management procedures and motivating consumer involvement in recycling initiatives. The overall effectiveness of EPR resides in its capacity to transfer the cost of waste management onto producers, encourage sustainable consumption patterns, and accelerate the achievement of environmental goals within the framework of a circular economy, despite obstacles like regulatory compliance and funding sources.

To mitigate the negative environmental effects of plastic trash, Extended Producer Responsibility (EPR) for plastic bottles has shown to be quite successful. EPR programs encourage advancements in recycling technologies and infrastructure by shifting accountability for the collecting, recycling, and disposal of plastic bottles to producers. Plastic bottle waste in landfills, the ocean, and natural habitats has decreased as a result of this, and recycling rates have improved dramatically. EPR promotes the use of recycled content and favours recyclable materials, encouraging businesses to innovate in packaging design. Additionally, education initiatives have increased consumer knowledge and involvement in recycling programs, which has helped to foster a sustainable culture. Economically, EPR lowers waste management and environmental remediation expenses while promoting employment growth in the recycling industry. The overall effectiveness of EPR for plastic bottles resides in its ability to drive systemic change towards a circular economy, where resources are used more efficiently and waste is minimized, thereby mitigating environmental harm and promoting long-term sustainability, even though challenges such as ensuring compliance and harmonizing EPR schemes across different regions still exist.

Although EPR can be very beneficial in reducing environmental harm, there are a number of issues with it that make it less effective in reaching sustainable waste management targets for the disposal of plastic bottles. Plastic bottle designs and materials are diverse and sophisticated, which can make recycling procedures more difficult. This is one major difficulty. Contemporary technologies find it challenging to recycle plastic bottles since they frequently comprise multiple types of plastic polymers, have multi-layered architectures, or contain mixed components. The diversity of materials decreases recycling operations' efficiency and raises processing costs. Furthermore, EPR plans necessitate large infrastructure investments for facilities for collecting, sorting, and recycling, which might not be feasible for all plastic bottle types or in all regions. Moreover, obstacles including inconsistencies in recycling labelling, convenience barriers, and low knowledge continue to make it difficult to attain significant levels of consumer engagement in recycling programs. To optimize the environmental advantages of Extended Producer Responsibility (EPR) for the disposal of plastic bottles, these constraints highlight the necessity of sustained innovation in recycling technologies, standardization of EPR regulations amongst jurisdictions, and improved consumer education. EPR regulation is subject to regular modifications, which can be challenging for compliance teams to stay up to date. Because of this, manufacturers find it difficult to comply with the latest reporting requirements and run the danger of not doing so.

In the future, Extended Producer Responsibility for plastic bottles has enormous potential to advance environmentally friendly waste management techniques and accomplish sustainable waste management objectives. Global awareness of plastic pollution is increasing, and EPR is well-positioned to lead the way in developing novel solutions that improve recycling rates and cut down on plastic waste. Improved recycling technology that can process a variety of plastic bottle materials more effectively in the future can help overcome the challenges we currently face in recycling intricate or multi-layered plastic designs. EPR frameworks may also be expanded to include stronger guidelines for product design, which would incentivize manufacturers to give priority to recyclability and use recycled content. Developing standardized EPR laws that support uniformity across regions and enable smooth recycling operations would require increased cooperation amongst stakeholders, including producers, governments, recyclers, and consumers. As we look to the future, EPR for plastic bottles will be most successful when it promotes a circular economy, in which materials are continuously recycled and reused, reducing their negative effects on the environment and guaranteeing future generations have a sustainable future [46, 47].

CHAPTER 4.3: LEGISLATIONS REGARDING BIOPOLYMERS AS BOTTLE PACKAGING

As environmentally friendly substitutes for conventional plastic bottles, biopolymers are drawing more and more attention as viable options for bottle packaging. Biopolymers are a feasible solution to reduce the pollution caused by plastic waste because they are derived from sources of renewable biomass like plants, algae, or microorganisms. These materials have qualities like flexibility, toughness, and moisture and oxygen barrier capabilities that make them ideal for packaging applications. Biopolymers are a major invention in the search for environmentally conscious packaging materials that are in line with the ideas of environmental stewardship and the circular economy, as demand for eco-friendly packaging increases. Their use in bottle packaging helps the packaging sector become more environmentally conscious and sustainable while also reducing its need on fossil fuels.

For the sake of maintaining environmental sustainability and public health, laws controlling the use of biopolymers in water bottles are essential. Customers are shielded from possible hazards related to chemical leak or contamination by these laws, which guarantee that biopolymer products adhere to strict safety standards. Laws that establish unambiguous standards and mandate testing encourage consumers to have faith in the dependability and quality of bottles made of biopolymer, which encourages their broad market acceptance. Furthermore, environmental concerns are of the utmost importance. Biodegradability, compostability, and recyclability criteria are frequently required by law, which lessens the environmental impact of packing materials and encourages appropriate waste management techniques. By promoting innovation in green technology, compliance with these standards not only advances the aims of sustainable development, but it also improves accountability and transparency throughout the supply chain. In the end, strong legal frameworks promote industry innovation towards a more sustainable future for bottle packaging while simultaneously maintaining product integrity and consumer confidence.

In the **United States**, FDA is principally responsible for overseeing rules pertaining to the usage of biopolymers in water packaging. Biopolymers that are meant to be used in water packaging either require a Food Contact Notification (FCN) or are the focus of an FDA-submitted food additive petition. In order to complete this process, data proving the biopolymer's safety for its intended application in contact with water must be submitted. Title 21 of the Code of Federal Regulations (21 CFR) governs the use of biopolymers in water packaging. More specifically:

- General rules for food additives, including definitions, safety standards, and exceptions, are found in 21 CFR Part 170.
- Indirect food additives, including polymers, for use on surfaces in contact with food are covered by 21 CFR Part 176. This rule specifies the conditions for safety assessments, which include assessing the possibility of material migration into water.

The FDA mandates safety evaluations to guarantee that biopolymers don't leak contaminants into water or endanger human health. Data on temperature, duration of contact, and types of water that demonstrate the material's safety under intended use conditions must be provided by the manufacturer. To guarantee that packaging materials are produced, processed, and handled in a safe and hygienic manner, producers of biopolymers are required to abide by the FDA's Good Manufacturing Practices (GMP) requirements (21 CFR Part 110). Additionally, biopolymers used in water packaging have to adhere to all applicable standards on environmental claims and recyclability, as well as the FDA's labelling requirements (21 CFR Part 101), which include accurate material identification. Although the FDA is primarily concerned with food safety, environmental impact studies for biopolymers are also taken into account. Although precise rules in this area are generally maintained by other organizations like the Environmental Protection Agency (EPA), this may involve assessments of biodegradability, compostability, or environmental safety considerations.

The **European Union** has extensive legislation and regulations that control the use of biopolymers in water packaging with the goal of maintaining environmental sustainability and safety. General guidelines for objects and materials meant to come into contact with food, drink, and water are established by Regulation (EC) No 1935/2004. To guarantee that they do not introduce dangerous materials into the water, biopolymers used in water packaging must adhere to strict migration limitations and safety regulations. Plastic materials and items meant to come into contact with food, including water, are specifically covered under Regulation (EU) No 10/2011. To prove conformity with the migration restrictions and safety criteria outlined in this rule, biopolymers used in water bottles must undergo stringent testing and certification procedures. Standards for compostability and biodegradability, such as those found in EN 13432 (a European standard that outlines the requirements for packaging recoverable by composting and biodegradation), may apply to biopolymers used in water packaging. By ensuring that biopolymers decompose naturally into innocuous molecules, these standards lessen their negative effects on the environment. Requirements for managing packaging waste, including biopolymers, are outlined in the EU Waste Framework Directive (2008/98/EC). With the goal of minimizing the environmental impact of packaging materials, it encourages waste prevention, recycling, and recovery. Finally, the EU's Circular Economy Action Plan seeks to advance sustainable patterns of production and consumption, including the use of compostable and biodegradable materials in packaging. In order to increase recyclable content and decrease waste production, it promotes creativity in packaging design.

In **India**, food packaging legislations and regulations are primarily governed by several key agencies and bodies that collaborate to ensure food safety, consumer protection, and environmental sustainability. Food Safety and Standards Authority of India (FSSAI) is a premier regulatory agency within the Ministry of Health and Family Welfare (MoHFW) of the Government of India. It is in charge of developing and implementing food safety standards, including laws pertaining to food packaging materials. The MoHFW is in charge of India's health-related laws and policies. It works with FSSAI to make sure food packaging laws comply with public health goals and requirements. The Ministry of Environment, Forest and Climate Change (MoEFCC) is in charge of sustainable development. In order to reduce its negative effects on the environment, it regulates packaging materials and encourages the use of eco-friendly and biodegradable materials. The MoEFCC oversees the Central Pollution Control Board (CPCB), which is tasked with preventing and reducing pollution. It might be involved in packaging material regulations to control their environmental impact, especially with regard to recycling and waste management. BIS creates standards and specifications for a range of goods and materials, including packaging materials, as was previously mentioned. It sets standards and requirements for quality that packaging materials must adhere to in order to guarantee product integrity and consumer safety. Lastly, the Food Safety Authority of each state in India is responsible for enforcing the FSSAI laws at the state level. Food safety regulations, particularly those pertaining to packaging materials used within their authority, are enforced by state authorities.

Together, these organizations create, carry out, and uphold laws pertaining to food packaging materials in India. In addition to reducing their negative effects on the environment, they work to protect public health, advance sustainable practices, and guarantee the safety of packaged foods. While there are numerous regulations and rules governing the use of plastics in food and consumables, these organizations have also established a few laws governing the use of biopolymers in food or liquid packaging. There are a number of laws and rules that guarantee food safety, environmental sustainability, and consumer protection apply to the use of biopolymers as bottle packaging.

1. **Food Safety and Standards (Packaging) Regulations, 2018**: These regulations by the FSSAI set out requirements for packaging materials used in direct contact with food. They specify permissible substances, overall migration limits (the number of substances that can migrate from the packaging to the food), and other safety criteria to prevent contamination of food by packaging materials, including biopolymers.

2. **Biodegradable Plastics Notification, 2012**: In this notice, the MoFCC lays forth the requirements for classifying plastics as biodegradable. It describes biodegradability, compostability, and ecotoxicity requirements that may apply to specific kinds of biopolymers used in bottles and other packaging materials.

3. **Plastic Waste Management Rules, 2016**: The production, use, recycling, and disposal of plastic items are governed by these regulations, which were later revised in 2018 and 2021. Instead of using traditional plastics, they promote the use of biodegradable and compostable materials, such as biopolymers, which are environmentally acceptable alternatives.

4. **Bureau of Indian Standards (BIS) Standard**: For a range of packaging materials, including biodegradable plastics, BIS has created standards and requirements. By adhering to these guidelines, biopolymers used in bottle packaging are guaranteed to satisfy strict quality and safety specifications.

5. **Environmental Protection Act, 1986**: The laws that govern pollution management and environmental protection in India are outlined in this comprehensive legislation. In order to lessen the influence on the environment, it emphasizes the significance of utilizing sustainable materials and successfully managing trash, which includes encouraging the use of biopolymers in packaging.

The development of biopolymers and other sustainable packaging solutions is actively encouraged by the Indian government and a number of research institutes. The goals of initiatives are to enhance the functionality, economic viability, and environmental impact of packaging materials made from biopolymers. Due to strict regulations and an increasing awareness of environmental sustainability, biopolymers have a bright future in bottle packaging in India. Biopolymers provide a biodegradable and compostable substitute for conventional plastics. They are made from renewable resources such as waste biomass or plant-based components. Their capacity to lower carbon emissions and plastic waste is in line with India's objectives for sustainable growth and environmental preservation. Biopolymers are anticipated to become increasingly affordable, adaptable, and scalable for usage in a range of packaging applications, including bottles, as science and technology progress. The usage of biopolymers in bottle packaging is anticipated to spread across sectors due to continuing advancements in material science and rising customer demand for environmentally friendly products. This will help India's packaging needs to have a more sustainable and greener future.

Future Scope and Possibilities

5

CHAPTER 5.1: SUSTAINABLE ALTERNATIVES

One of the most frequent items of waste that litter our surroundings are plastic bottles. A single bottle's carbon footprint during production ranges from 300 to 500 g carbon dioxide. Bottles contaminate our oceans, fill up landfills, and may even put wildlife in danger. Plastic is a convenient material to use for carrying liquids, such as soda and water, but aside from producing carbon dioxide, it has a significant drawback: it often ends up as trash. It is estimated that 481 billion plastic water bottles are used globally, and fewer than 30% of those bottles are recycled. What's left either ends up as pollution in our streams and oceans or in landfills where it can take generations for them to break down. For these reasons, it's imperative to find plastic bottle alternatives. Primarily composed of petroleum-derived polyethylene terephthalate (PET), conventional plastic bottles are a major source of environmental contamination. Numerous of them wind up in seas, where they damage ecosystems and marine life over the course of hundreds of years to degrade. To further exacerbate climate change, the manufacturing of plastic bottles releases greenhouse gas emissions and uses fossil fuels. Recycling, composting, and biodegradable alternatives are available, such as bioplastics, glass, aluminium, and plant-based materials. These substitutes alleviate environmental effects, encourage circular economy principles, and lessen dependency on finite fossil resources. Additionally, they meet the growing need for eco-friendly products and customer awareness, which pushes industry to take

 DOI: 10.4324/9780000000002-5

FIGURE 5.1   Glass bottles

innovative approaches to more sustainable packaging. Globally, including in India, a number of environmentally friendly alternatives to conventional plastic bottles are being investigated and put into practice. Listed below are a few promising choices:

a. Glass bottles: For a number of strong reasons, glass bottles are generally acknowledged as being better and more environmentally friendly than plastic bottles (Figure 5.1). Initially, glass is naturally environmentally favourable since it can be recycled indefinitely without losing its quality, unlike plastic that deteriorates after a few recycling cycles. Because of its recyclability, bottles may be made with less energy and raw materials, which helps with conservation efforts and lowers carbon footprints.

Additionally, because glass is inert and does not absorb dangerous substances, drinks or other goods kept inside are safe and pure. Glass bottles are especially good for products that need to be stored for a long time or are susceptible to contamination because of this feature. Glass can also be recycled multiple times and is extremely robust, which encourages the circular economy of reusing things instead of discarding them after only one usage. Glass's reputation as a premium material also contributes to its consumer appeal and fits in with consumers' increased demands for premium, eco-friendly packaging options. Overall, glass bottles are a preferred option for companies and customers who are dedicated to limiting their environmental effect and eliminating plastic waste because of its safety, durability, and recyclability.

While using glass bottles has benefits, there are a number of serious drawbacks, especially when considering practicality and environmental impact. First off, glass bottles are far heavier than their plastic counterparts, which raises the cost of shipping and increases carbon emissions. The fragility of glass paired with the additional weight increases the likelihood that it may break during handling, transportation, and storage, increasing product loss and safety risks. In addition, the manufacturing of glass bottles necessitates elevated temperatures, leading to significant energy usage and an increased carbon footprint in contrast to the production of plastic, particularly when the glass is not derived from recycled resources. Although glass recycling has many benefits, there are drawbacks to the process as well. Colour sorting and contamination can make recycling more difficult and reduce the value of the material. Furthermore, compared to plastic bottles, glass bottles have less design versatility, which could limit branding and product uniqueness. The heavier and more delicate nature of glass means that handling and storage become more difficult, requiring more expensive and sophisticated solutions. Heavy, breakable containers can be inconvenient for customers, especially when it comes to outdoor or on-the-go activities. Despite the fact that glass is inert and does not absorb dangerous substances, glass bottles that have been disposed of incorrectly can linger in the environment and cause litter and safety risks. Businesses thinking about using glass bottles for packaging must weigh these drawbacks against their aesthetic appeal and capacity to be recycled.

b. Bioplastics: Bioplastics are a major development in environmentally friendly packaging, providing a number of benefits over conventional plastic bottles. Bioplastics are made from renewable biomass sources including cellulose, sugarcane, or maize starch. Depending on their composition, they can be composted, biodegradable, or both. Because bioplastics may naturally decompose under composting conditions, they lessen the impact that plastic waste has on landfills and marine environments. This feature tackles one of the main environmental problems related to plastic waste. Additionally, compared to traditional plastics, the manufacture of bioplastics often uses less fossil fuels, which lowers greenhouse gas emissions and lessens dependency on limited resources. Additionally adaptable for use in bottles and other packaging applications, bioplastics can be engineered to share attributes with conventional plastics, such as strength and barrier qualities. Furthermore, the use of bioplastics boosts the agricultural industry by opening up new markets for biomass resources, which in turn encourages rural development and improves economic sustainability. Bioplastics are a promising option for companies and consumers looking for environmentally friendly alternatives to plastic bottles, even though issues like cost-effectiveness and scalability still exist. This is because bioplastics are becoming more and more viable and effective thanks to ongoing research and technological advancements.

A wider variety of packaging options are made possible by the distinct benefits that various types of bioplastics—which are used to make bottles—offer. A well-liked bioplastic with outstanding visual appeal, polylactic acid (PLA) is made from sugarcane or maize starch and is characterized by its rigidity and clarity, which make it ideal for transparent bottles. Additionally biodegradable in industrial composting settings, PLA lessens its long-term environmental impact. The byproduct of microbial fermentation of sugars, polyhydroxyalkanoates (PHA) are excellent for bottles that might wind up as litter because of their adaptability and biodegradability in a variety of settings, including soil and seawater. The better barrier qualities of plant-based sugars, known as polyethylene furanoate (PEF), provide it an advantage over traditional polyethylene (PET) when it comes to protecting perishable goods like beverages from gases like carbon dioxide and oxygen. In addition, PEF is less carbon intensive and may be recycled using the current PET recycling channels. Along with lowering greenhouse gas emissions and dependency on fossil fuels, these various bioplastics also provide unique functional advantages such enhanced barrier qualities, biodegradability, and aesthetic versatility. A useful substitute for conventional plastics, the use of various bioplastics in the bottle-making process enables customized solutions that satisfy particular product requirements and environmental aims.

c. Aluminium: There are various environmental and practical reasons why aluminium bottles are an appealing replacement for conventional plastic bottles. First of all, aluminium is a highly recyclable material that can be recycled endlessly without losing its quality. Because of its recyclability, less energy is used during production than during the extraction of virgin aluminium, which dramatically decreases the need for new raw materials. Aluminium recycling is a more environmentally friendly option overall since it can save up to 95% of the energy used in generating aluminium from bauxite ore.

Aluminium bottles also have exceptional durability and damage resistance due to their strength and low weight. The quality and freshness of drinks or other liquids kept within are preserved by their efficient barrier qualities against light, air, and moisture as well as their impermeability. Aluminium bottles can also be reused, which supports the circular economy's idea of reusing things instead of discarding them after just one use. For consumers and organizations who want to reduce their environmental impact while maintaining product purity and customer happiness, aluminium bottles are a great choice because of these features. Aluminium bottles are positioned to become a more important component of sustainable packaging solutions globally as companies continue to innovate and improve recycling techniques.

While there are certain advantages to using aluminium bottles, there are also some significant drawbacks that affect sustainability and practicality. The extraction and processing of bauxite ore, which is required in the highly energy-intensive process of producing aluminium, greatly increases greenhouse gas emissions and degrades the environment. Aluminium can be recycled, however there is a significant carbon impact in the initial production process. Aluminium bottles can also be easily dented and deformed, which reduces their strength and visual appeal. Despite not being as brittle as glass, they can nevertheless get worn out, especially after prolonged use. Leaching is a further issue as, in the absence of a suitable lining, metal can react with some drinks and compromise their safety and flavour (Figure 5.2).

Due to its reactive nature, aluminium may undergo chemical reactions that could have an impact on the contents when it comes into touch with acidic or alkaline substances. Fruit juices, drinks with carbonation, and some alcoholic beverages, for example, are acidic and may leach metal into the liquid, changing its flavour and possibly posing health risks. An inert liner, usually composed of epoxy or polymer resins, is frequently applied to aluminium bottles to help reduce this effect by acting as a barrier across the metal and the liquid inside. Direct contact and potential contamination could result from the protective barrier failing if this lining becomes harmed or degrades over time. Thus, even though aluminium bottles have benefits like being recyclable and lightweight, maintaining the integrity of the inside coating is essential to avert negative reactions and preserve the security and calibre of the goods being held. Furthermore, aluminium bottles may not be as economically feasible to produce on a wide scale as plastic bottles due to their

FIGURE 5.2 Aluminium bottle

higher production costs. Despite their low weight, which helps to save on transportation expenses, there are issues with their total environmental impact and material constraints. Finding the right aluminium bottle for packaging requires weighing these disadvantages against their lightweight design and capacity to be recycled.

d. Biodegradable disposable bottles: The environmental problems that standard plastic bottles cause could potentially be solved by biodegradable bottles. These bottles, which are composed of biodegradable materials like plant-based polymers or bioplastics, can spontaneously disintegrate into non-toxic components when they come into contact with heat, moisture, and microbes. One of the most important problems with plastic pollution is that this capacity greatly lessens the amount of plastic trash that remains in land-fills and marine habitats. When composted, biodegradable bottles enhance soil health by adding organic matter, as opposed to leaving behind dangerous microplastics.

Apart from its advantages for the environment, biodegradable bottles can also be made from renewable resources like sugarcane or corn starch, which lessens reliance on finite fossil fuels and the greenhouse gas emissions linked to the production of conventional plastic. Biodegradable polymers are more sustainable than traditional plastics since their production procedures usually need less energy and water. Furthermore, biodegradable bottles can continue to function and last as long as conventional plastic bottles, making them appropriate for a variety of uses, such as drinks and personal hygiene items. Biodegradable bottles are a practical choice for companies looking to meet sustainability targets and satisfy customers who care about the environment as consumer awareness of and demand for eco-friendly substitutes rise.

Even if there are still issues with scalability and cost-effectiveness, research and development in technology are improving the availability and performance of biodegradable materials. Biodegradable bottles offer a promising option to cut plastic pollution and promote a more sustainable future for packaging solutions worldwide when used in conjunction with appropriate waste management systems.

e. Paperboard bottles: Paperboard bottles have a number of significant advantages over typical plastic bottles that make them attractive sustainable alternatives. Paperboard bottles are less harmful to the environment than plastic bottles since they are made of renewable and biodegradable materials, such as paper fibres from forests that are ethically managed or agricultural wastes. Because of their recyclable and biodegradable nature, they offer a variety of end-of-life solutions that help cut waste and advance the ideas of the circular economy.

Paperboard bottles are strong and lightweight, providing enough strength to hold liquids such as milk, juice, and other drinks. To make sure they don't leak and keep the goods fresh, they might be made with coatings or inner linings. Depending on how they are made and what other coatings are put on them, these bottles can also have effective light, oxygen, and moisture barriers. Additionally, compared to the production of plastic bottles, the creation of paperboard bottles often uses less energy and water. They support initiatives aimed at mitigating climate change and environmental deterioration by helping to reduce greenhouse gas emissions and dependency on fossil fuels.

Additionally, paperboard is a material that is widely accepted for recycling in many areas, which helps to divert waste from landfills and increase recycling rates.

Paperboard bottles appeal to consumers as natural and environmentally friendly, which is in line with their growing desire for eco-friendly packaging options. They can be personalized with printing and branding options, making them appropriate for a range of liquid items while preserving the product's marketability and visibility.

f. Bio-based PET bottles: PET bottles made of biodegradable materials represent a substantial advancement in packaging sustainability. PET bottles constructed from renewable biomass sources, like sugarcane or maize starch, have a lower dependency on finite fossil fuels than typical PET bottles derived from petroleum. Throughout the course of the product lifecycle, this renewable sourcing helps reduce greenhouse gas emissions and its carbon footprint. PET bottles made from biomass share many of the same characteristics as regular PET bottles, making them strong, lightweight, and adaptable to a wide range of packaging uses. They continue to have strong carbon and oxygen barrier qualities, which is essential for keeping drinks and other liquid goods fresh and high-quality.

Furthermore, bio-based PET bottles can be recycled in all current PET recycling programs. Along with traditional PET bottles, they can be gathered, sorted, and processed, promoting circular economy principles and minimizing trash buildup in landfills. Since recycled PET can be used to create new bottles or other products, extending their lifecycle, recycling bio-based PET bottles contributes to resource and energy conservation.

Bio-based PET bottles are a sustainable substitute that don't sacrifice usability or performance from the perspective of the consumer. They are comfortable and functional. They support brand identity and environmental stewardship while also meeting the growing demand from consumers for eco-friendly packaging options. Even with the presence of obstacles like production scalability and cost competitiveness, bio-based PET bottle viability and efficiency are rising due to continuous research and technical improvements. Bio-based PET bottles are positioned to play a significant part in lowering plastic waste and moving towards a more sustainable future for packaging solutions globally, thanks to rising environmental awareness and governmental support for sustainable practices.

g. Silicone bottles: When it comes to longevity, reusability, and environmental impact, silicone bottles present a strong case as environmentally friendly substitutes for conventional plastic bottles.

For a variety of uses, including the storage of food and beverages, silicone is a versatile and long-lasting material that can tolerate both high and low temperatures. Longer product lifespans result from this endurance, which also minimizes waste production and the need for frequent replacements.

In addition, unlike certain plastics, silicone bottles are reusable and can tolerate repeated use without deteriorating or leaking dangerous substances into contents. By encouraging longer product lifecycles and minimizing waste from single-use plastics, this reusability not only reduces overall resource consumption but also advances a circular economy.

Silicone is a safe option for food and drink storage because it is naturally non-toxic, hypoallergenic, and resistant to the formation of bacteria. Because of its stability and non-reactive nature, it preserves the integrity of the product and the delight of the customer by not changing the flavour or quality of the liquids it holds.

Silicone is thought to be more environmentally friendly than conventional polymers. In certain areas with dedicated recycling programs, it can be recycled even though it is not biodegradable. Lower environmental impact is also achieved via silicone manufacturing procedures, which generally use less energy and emit fewer emissions than those used in the production of traditional plastics. All things considered, silicone bottles provide a safe, reusable, and long-lasting substitute for plastic bottles. In an effort to lessen plastic pollution and encourage sustainable living, silicone bottles are projected to become more widely available in the market as consumer awareness and demand for eco-friendly products rise.

h. Ceramic bottles: With its unique features, ceramic bottles are an attractive and environmentally friendly substitute for conventional plastic bottles. Primarily, ceramic is a naturally occurring, non-toxic substance that doesn't contaminate drinks with hazardous substances, guaranteeing the safety and integrity of the liquids kept inside. For consumers looking for a more healthful option than plastic bottles, ceramic bottles are a great option. Furthermore, because of their innate toughness and resilience to deterioration, ceramic bottles have a lengthy lifespan. Reusable again and time again without sacrificing quality, they are less prone to dents, scratches, and degradation than plastic. Because there is less need for regular repairs and less waste produced, its durability encourages sustainability.

Additionally, ceramic bottles can have a beautiful appearance. They provide a range of design options with vivid colours and detailed patterns that accentuate their visual attractiveness. Ceramic bottles have inherent material properties and are particularly useful for designing packaging that is both visually appealing and adaptable. Unique and complex patterns that stick out on the shelves of shops can be created by moulding ceramics into a range of shapes and sizes. What makes the material even more visually appealing is its capacity to retain intricate patterns and vivid colours through coating and painting techniques. Additionally, compared to plastic or metal substitutes, ceramic bottles have a natural, premium feel and appearance that is frequently seen as more opulent and handmade. This elegant design can draw in customers searching for unique, fashionable packaging and greatly improve your image. Ceramic bottles also have great insulating qualities that assist prolong the temperature of liquids without requiring extra energy from heating or cooling appliances. Comparing this energy efficiency to plastic bottles results in a reduced total environmental effect. Ceramic bottles are visually appealing to consumers and are available in a range of patterns and styles, accommodating personal tastes and improving the whole drinking experience. They also encourage sustainable living habits and lessen the environmental impact of single-use plastics.

As efforts to reduce plastic pollution gain momentum and worldwide awareness increase, the market for sustainable bottles is expected to see significant expansion and innovation. Important patterns point to a move towards more sophisticated and environmentally friendly materials, like bioplastics made from sugarcane or corn starch, which are renewable resources, and bio-based plastics, which lessen reliance on fossil fuels. In addition to being recyclable or biodegradable, these materials also fulfil performance standards that are on par with those of typical plastics, meaning they can be used in a variety of industries, from beverages to personal care items.

CHAPTER 5.2: ISSUES TO BE ADDRESSED IN ADVANCED R&D IN PLASTIC BOTTLE RECYCLING TECHNOLOGIES

In order to manage the growing global plastic waste challenge and make the shift to a circular economy, advanced research and development (R&D) in plastic bottle recycling technology is essential. Traditional recycling techniques have advanced significantly, but there are still a number of serious issues that prevent them from being more effective, scalable, and environmentally sustainable. These obstacles include but are not limited to technological constraints, financial feasibility, legal and regulatory environments, and the requirement for creative ways to improve material recovery and quality. Through cutting-edge research and development, stakeholders hope to address these problems and transform the recycling of plastic bottles, lessen their negative effects on the environment, and encourage sustainable business practices. To increase effectiveness, scalability, and environmental impact, advanced R&D in plastic bottle recycling techniques must solve a number of significant obstacles.

a. Contamination and sorting: In the field of advanced research and development for plastic bottle recycling systems, contamination and sorting are critical topics. The quality and purity of recycled materials are severely hampered by contamination from leftover liquids, labels, mixed plastic kinds, and other non-plastic elements. The mechanical and chemical qualities of recycled plastics may be jeopardized by these contaminants, which could reduce their value and usefulness in later production processes. Therefore, overcoming these obstacles requires the use of advanced sorting methods. The efficiency and precision of sorting are being improved by innovations like optical sorting systems with sensors and artificial intelligence (AI). Greater purity levels in the recycled streams are ensured by these technologies, which allow for the accurate identification and separation of various plastic kinds and pollutants.

Advanced sorting optimizes resource usage and minimizes waste, which not only improves the quality of recovered plastics but also lowers operational costs, increases process efficiency, and supports environmental sustainability. Moreover, efficient contamination control promotes market acceptability of recycled materials and helps enterprises comply with strict regulations, which propels the development of more environmentally friendly packaging options. To put it simply, improving sorting technologies and tackling contamination through research and development is essential to increasing the effectiveness and scalability of recycling plastic bottles and opening the door to a more sustainable and circular economy [48].

b. Mechanical recycling challenges: One major area of focus for research and development efforts to improve plastic bottle recycling systems is mechanical recycling problems. In order to create recycled resin pellets or flakes, conventional mechanical recycling methods include shredding, cleaning, and melting plastic bottles. These procedures, however, confront a number of significant obstacles that reduce their efficacy and efficiency. First off, a problem with plastic trash is its complexity because it frequently consists of several kinds of polymers with unique chemical compositions and physical characteristics. Accurately sorting and separating these polymers is necessary to guarantee that only elements that are suitable are processed together, reducing contamination and maintaining the quality of the recycled material. To attain improved accuracy and throughput in sorting, advanced R&D focuses on enhancing sorting technologies such optical sorting systems, near-infrared (NIR) spectroscopy, and AI-driven algorithms.

Second, successfully preserving the mechanical and chemical characteristics of recycled plastics is frequently out of reach for mechanical recycling techniques. Reduced material strength, durability, and thermal stability can be the result of polymer chains being broken down by repeated heating and shredding cycles. This deterioration restricts the use of recycled plastics in high-end items in addition to impairing their performance. The goal of advanced research and development is to create compatibilizers, additives, and processing methods that will slow down polymer deterioration and improve the quality and functionality of recycled materials.

Another important factor to take into account is how to increase mechanical recycling processes while keeping them economically viable. In order to increase the mechanical recycling technologies' economic feasibility, research activities are focused on process efficiency optimization, energy consumption reduction, and operational workflow streamlining. The objectives of the circular economy, which include minimizing plastic waste and optimizing resource efficiency, depend on these developments in order to encourage the broad use of sustainable recycling techniques.

c. Chemical recycling: By addressing crucial issues that mechanical recycling is unable to resolve, chemical recycling research and development (R&D) holds great potential for revolutionizing the recycling of plastic bottles. polymers are broken down into their molecular components, such as monomers or feedstock chemicals, using processes like depolymerization or pyrolysis in chemical recycling techniques as opposed to mechanical recycling, which includes melting and reforming polymers. More types of plastics, including polluted or mixed materials that are challenging to treat mechanically, can be recycled because to this capability.

Additionally, chemical recycling has the potential to yield high-purity recycled materials that are on par with virgin plastics, increasing the market opportunity and range of applications for recovered plastics. Comparing chemical recycling technologies to more conventional recycling techniques, they also seek to increase overall environmental sustainability, lower greenhouse gas emissions, and improve energy efficiency. Recent studies indicate that current research and development efforts are concentrated on streamlining these procedures to attain increased yields, reduced expenses, and increased scalability, thus quickening the shift in the plastics recycling industry towards a more circular economy.

1. Depolymerization: Plastic polymers are broken down into their basic monomer components using a chemical recycling process called depolymerization of plastic bottles. By efficiently reversing the polymerization process, this technique turns complex plastic materials into simpler, reusable materials by the use of heat, catalysts, or chemical agents. For example, the plastic bottle material polyethylene terephthalate (PET) can be depolymerized back into its original monomers, ethylene glycol and terephthalic acid. In order to create new PET and preserve the material's excellent quality and functionality, these monomers can subsequently be cleaned and repolymerized. This method has a number of benefits, one of which is its capacity to handle contaminated or mixed plastics that would be challenging to mechanically recycle. Furthermore, depolymerization facilitates the production of superior recycled polymers that can be incorporated into new goods, thereby completing the circle in plastic recycling. Nevertheless, in order to grow efficiently, the process needs sophisticated equipment and infrastructure, which might be energy-intensive. Despite these obstacles, depolymerization is a viable way to improve plastic recycling and support a circular economy that is more sustainable.

2. Pyrolysis: Plastic bottles can undergo pyrolysis, a thermal breakdown process that uses high temperatures and no oxygen to break down plastic polymers into smaller molecules. Plastic trash is heated in this process to temperatures that range from 350°C to 700°C, which causes the waste to break down into a mixture of gases, liquids, and solid wastes. The gaseous byproducts, which are mainly hydrocarbons, can condense into liquid fuels like gasoline or diesel, and the solid residue, or char, can be utilized in other ways or as a carbon-rich material. A variety of plastic kinds, including contaminated and mixed plastics, which are frequently inappropriate for mechanical recycling

techniques, can be handled using pyrolysis. Moreover, it can recover important chemicals or energy and aid in the reduction of plastic waste volume. To run efficiently, the procedure necessitates a large infrastructure and a lot of equipment and might be energy-intensive. Notwithstanding these difficulties, pyrolysis is a promising technique that can help to create a more sustainable waste management system by turning plastic trash into products that are valuable.

The goals of current research and development initiatives pertaining to plastic bottle pyrolysis are to increase the method's effectiveness, sustainability, and adaptability. The goal of recent developments is to improve the yield and quality of the final products, which include gases, liquids, and solid wastes, by optimizing pyrolysis parameters like temperature and pressure. With a focus on producing high-value chemicals and fuels, researchers are investigating the use of sophisticated catalysts to speed up pyrolysis reactions and enhance product creation selectivity. Creating pyrolysis systems that are more energy-efficient and combining them with other waste-to-energy technologies is also a hot topic in an effort to enhance total energy recovery. Pyrolysis reactor design and operation are evolving to accommodate a wider variety of plastic types, including contaminated and mixed plastics.

3. Solvolysis: Using solvents to dissolve plastic polymers into their component monomers or other useful chemical components is a chemical recycling process called solvolysis of plastic bottles. By dissolving plastics in particular solvents, the polymer chains are selectively interacting with the solvents, cleaving them into smaller, more controllable molecules. For instance, by employing solvents that efficiently dissolve the polymer while removing impurities, solvolysis can be used to break down polyethylene terephthalate into its monomers, terephthalic acid and ethylene glycol. It is possible to treat a wide range of polymers using this approach, and it also recovers high-purity monomers that can be used to create new products by repolymerization. In addition to helping finish the recycling loop by supplying premium feedstocks for production, solvolysis is especially helpful for treating polymers that are challenging to recycle mechanically. To make solvolysis a more viable and sustainable recycling option, certain obstacles must be overcome, including the requirement for effective solvent recovery, the high cost of solvents, and the handling of solvent waste.

In the field of plastic bottle solvolysis, research and development trends are focused on improving the process's economic viability, efficiency, and selectivity. In order to dissolve plastic polymers like polyethylene terephthalate (PET) selectively without harming other materials or producing an excessive amount of trash, recent advances have concentrated on creating and refining solvents. To reduce the environmental impact and operating expenses of the solvolysis process, researchers are investigating new solvent systems, such as green solvents and mixes. Refining reaction parameters like temperature, pressure, and solvent concentration is increasingly gaining popularity as a way to optimize the conversion of plastics into useful monomers or chemicals. Catalyst development innovations are being explored to enhance the recovery of high-purity monomers and speed up solvolysis operations.

4. Gasification: The method of gasification involves heating plastic bottles to a high temperature and using steam reforming or partial oxidation with a small amount of oxygen to create syngas, a synthetic gas. Heat treatment, usually between 700°C and 1,200°C, is applied to plastic waste during gasification. This causes the trash to break down into a combination of gases, mostly carbon dioxide, hydrogen, and carbon monoxide. Both energy production and chemical synthesis, including the synthesis of methanol and ammonia, can be supported by this syngas.

Enhancing the process's economy, sustainability, and efficiency are the main goals of current research and development trends in plastic bottle gasification. By enhancing reaction conditions and technology, recent developments seek to enhance the conversion of plastic waste into high-quality synthetic gas, or syngas. The creation of sophisticated catalytic materials that improve gasification reactions and increase the yields of valuable gases like carbon monoxide and hydrogen is one of the major trends. To improve energy efficiency and lower the process's overall energy consumption, researchers are also looking into novel reactor designs and operating scenarios. To optimize energy recovery and utilization, there is also rising interest in combining gasification with other waste-to-energy technologies, such as combined heat and power (CHP) systems. In addition, developments in syngas conditioning and cleaning technologies are underway to eliminate contaminants and make syngas more usable for a range of industrial uses.

By enhancing gasification's sustainability and viability as a plastic waste management technique, these R&D trends hope to lessen its negative environmental effects and promote a circular economy.

5. Hydrolysis: Plastics are hydrolysed (broken down into their monomer components) using catalysts and water. With hydrolysable plastics like PET, this method works especially well. Reclaimed high-quality monomers can be used in other applications or to make new plastic goods. The goal of plastic bottle hydrolysis trends is to maximize the chemical recycling process's sustainability, scalability, and efficiency. The focus of recent developments has been on increasing the efficiency of hydrolysis in dissolving plastic polymers, especially polyethylene terephthalate (PET), into their component monomers. In order to improve the rate and selectivity of hydrolysis reactions and increase the process's efficiency and viability from an economic standpoint, researchers are creating novel catalysts and refining reaction conditions. There is also a lot of interest in investigating more environmentally friendly and sustainable hydrolysis techniques, such breaking down plastics with bio-catalysts made of microorganisms or using water under extreme pressure and temperature. In order to enhance system performance and material recovery, advances are also being made in the integration of hydrolysis with other recycling processes. Moreover, efforts are made to make hydrolysis technologies more scalable so that industrial settings can utilize them to their full potential. The aforementioned developments are geared towards promoting hydrolysis as a crucial technique for generating superior recycled materials, bolstering the circular economy, and tackling the obstacles associated with managing plastic waste [49].

d. Biodegradable additives: By addressing issues with material compatibility and end-of-life management, biodegradable additives research and development provide a possible path to improve plastic bottle recycling. Biodegradable additives are engineered to make it simpler for plastics to break down during recycling procedures, hence increasing the effectiveness of mechanical and chemical recycling techniques. To improve plastic bottles' biodegradability or make it easier for them to separate from other plastics like polyethylene terephthalate (PET) and polypropylene (PP), which are frequently used in bottle production, these additives can be added to plastic bottles as they are being manufactured. When plastics reach the end of their useful lives, these additives lessen their environmental impact by improving their biodegradability, which allows them to break down more quickly in anaerobic digestion or composting facilities.

Furthermore, by reducing contamination in recycling streams, biodegradable additives can raise the calibre and purity of recovered materials. By making it possible for recycling procedures to be more effective and by expanding the supply of premium recycled plastics for new products, this supports the circular economy. Recent studies indicate that research and development (R&D) efforts are concentrated on enhancing the efficacy and suitability of biodegradable additives with diverse plastic kinds and assessing the environmental advantages of these additives in various recycling situations. With the goal of promoting sustainable practices and lowering plastic pollution in ecosystems, these initiatives seek to smoothly incorporate biodegradable chemicals into the current plastic bottle manufacturing processes [50].

e. Energy-efficient processing: By tackling the high energy consumption associated with existing recycling technologies, advanced research and development (R&D) in energy-efficient processing represents a critical option for enhancing the recycling of plastic bottles. Traditional recycling methods, such mechanical recycling, frequently need a lot of energy to melt, shred, and reform plastic waste. These procedures can be expensive to run on a large scale and add to the carbon footprint of recycling plants. The goal of advanced research and development in energy-efficient processing is to create novel tools and techniques that will lower the energy consumption of recycling plastic bottles. Examples of technology being investigated to improve the efficiency of plastic melting and shaping operations include microwave-assisted recycling and sophisticated thermal processing methods. These techniques seek to outperform traditional techniques in terms of throughput rates and energy usage, which will cut down on both operating expenses and environmental effect.

Incorporating renewable energy sources into recycling plants also improves sustainability since it minimizes greenhouse gas emissions related to recycling operations and lessens dependency on fossil fuels. Improved R&D in energy-efficient processing not only helps wider environmental aims but also increases the economic feasibility of recycling projects by improving energy efficiency throughout all stages of the recycling process, from collecting and sorting to reprocessing and manufacture.

Recent studies indicate that the main goals of current research and development (R&D) activities in energy-efficient processing are process parameter optimization, equipment design optimization, and lifetime environmental effect assessment of new technologies. These developments play a crucial role in encouraging the recycling of plastic bottles according to sustainable methods, promoting a more circular economy, and lowering resource consumption in general.

f. Biological recycling: Utilizing biological processes to degrade and recycle plastics, such as enzymatic degradation or microbial action, is known as biological recycling of plastic bottles. This strategy uses artificially created biological systems or natural organisms as a potential replacement for conventional mechanical and chemical recycling techniques in the management of plastic waste. The primary ideas and techniques related to biological recycling are as follows:

1. Enzymatic degradation: Certain enzymes have the ability to decompose polymeric polymers into their component monomers. Enzymes such as PETase and MHETase, for instance, specifically target PET, a common plastic used in bottles. PET is broken down into its intermediate product by PETase and further broken down into terephthalic acid and ethylene glycol by MHETase. To break down plastic into simpler molecules that can be recycled into new plastics or processed further, the procedure usually entails exposing the plastic to these enzymes under carefully regulated circumstances.

2. Microbial Degradation: Plastics can be broken down and used as a food source by some fungi, bacteria, and other microbes. As an illustration, it has been discovered that fungi like *Aspergillus* species and bacteria like *Ideonellasakaiensis* can destroy different kinds of plastics. Culture of these microorganisms in plastic waste-filled settings is necessary for microbial breakdown. Plastics can be further broken down into smaller, biodegradable molecules by the enzymes produced by the bacteria and then mineralized or further processed.

3. Combination of Enzymes and Microbes: Enzymes and bacteria work together to recycle plastic bottles, which is a clever and promising way to improve plastic waste management. This hybrid approach starts with enzymes that specifically target

and degrade complex plastic polymers, including polyethylene terephthalate (PET), into smaller, easier-to-handle molecules. For example, PET can be processed more easily by microorganisms when enzymes like PETase break it into its monomer components. Specialized microorganisms like bacteria and fungi further break down these smaller molecules once they have been broken down by enzymes, producing non-toxic byproducts like carbon dioxide and water. Combining the advantages of both biological processes, enzymes quicken plastics' early disintegration while microorganisms guarantee more thorough decomposition of the leftover pieces. By using such biological techniques, the impact on the environment and energy consumption may be reduced as there will be less reliance on the harsh chemicals and energy-intensive procedures that are typical of traditional recycling. The development of cost-effective techniques, maximizing the efficiency of microorganisms and enzymes, and scaling these processes for commercial usage continue to present difficulties. This integrated approach, in spite of these obstacles, has great potential to resolve the worldwide problem of plastic waste and advance sustainable recycling technology.

References

1. I. Gheorghe, P. Anastasiu, G. Mihaescu, and L.-M. Ditu, "Advanced Biodegradable Materials for Water and Beverages Packaging," in *Bottled and Packaged Water*, Grumezescu, A.M. and Holban, A.M. (eds.), vol. 4, Elsevier, 2019, pp. 227–239. doi: 10.1016/B978-0-12-815272-0.00009-X.
2. L. K. Ncube, A. U. Ude, E. N. Ogunmuyiwa, R. Zulkifli, and I. N. Beas, "Environmental Impact of Food Packaging Materials: A Review of Contemporary Development from Conventional Plastics to Polylactic Acid Based Materials," *Materials*, vol. 13, no. 21, p. 4994, Nov. 2020, doi: 10.3390/ma13214994.
3. L. Chen and Z. Lin, "Polyethylene: Properties, Production and Applications," in *2021 3rd International Academic Exchange Conference on Science and Technology Innovation (IAECST)*, IEEE, Guangzhou, China, Dec. 2021, pp. 1191–1196. doi: 10.1109/IAECST54258.2021.9695646.
4. Z. Yao, H. J. Seong, and Y.-S. Jang, "Environmental Toxicity and Decomposition of Polyethylene," *Ecotoxicology and Environmental Safety*, vol. 242, p. 113933, Sep. 2022, doi: 10.1016/j.ecoenv.2022.113933.
5. A. Emblem, "Plastics Properties for Packaging Materials," in *Packaging Technology*, Emblem, A. and Emblem, H. (eds.), Elsevier, 2012, pp. 287–309. doi: 10.1533/9780857095701.2.287.
6. Plast Reagent, "Understanding the Use and Need of HDPE Plastic Bottles," *Medium*, https://regentplast.com/making-of-chemical-hdpe-bottles-behind-the-scenes/.
7. J. L. Jordan, D. T. Casem, J. M. Bradley, A. K. Dwivedi, E. N. Brown, and C. W. Jordan, "Mechanical Properties of Low Density Polyethylene," *Journal of Dynamic Behavior of Materials*, vol. 2, no. 4, pp. 411–420, Dec. 2016, doi: 10.1007/s40870-016-0076-0.
8. M. M. S. M. Sabee, N. T. T. Uyen, N. Ahmad, and Z. A. A. Hamid, "Plastics Packaging for Pharmaceutical Products," in *Encyclopedia of Materials: Plastics and Polymers*, Elsevier, 2022, pp. 316–329. doi: 10.1016/B978-0-12-820352-1.00088-2.
9. Shell, *Polyethylene in Everyday Life: Common LLDPE Plastic Products*, Shell USA.
10. C. Maes, W. Luyten, G. Herremans, R. Peeters, R. Carleer, and M. Buntinx, "Recent Updates on the Barrier Properties of Ethylene Vinyl Alcohol Copolymer (EVOH): A Review," *Polymer Reviews*, vol. 58, no. 2, pp. 209–246, Apr. 2018, doi: 10.1080/15583724.2017.1394323.
11. K. K. Mokwena and J. Tang, "Ethylene Vinyl Alcohol: A Review of Barrier Properties for Packaging Shelf Stable Foods," *Critical Reviews in Food Science and Nutrition*, vol. 52, no. 7, pp. 640–650, Jul. 2012, doi: 10.1080/10408398.2010.504903.

12. H. Maddah, "Polypropylene as a Promising Plastic: A Review," —*American Journal of Polymer Science*, vol. 2016, pp. 1–11, Jan. 2016, https://www.researchgate.net/publication/290439450_Polypropylene_as_a_Promising_Plastic_A_Review.

13. I. Gheorghe, P. Anastasiu, G. Mihaescu, and L.-M. Ditu, "Advanced Biodegradable Materials for Water and Beverages Packaging," in *Bottled and Packaged Water*, Grumezescu, A.M. and Holban, A.M. (eds.), Elsevier, 2019, pp. 227–239. doi: 10.1016/B978-0-12-815272-0.00009-X.

14. Thomas Xometry, *Plastic Bottle Manufacturing Process – How Plastic Bottles Are Made*, Thomas A Xometry Publication.

15. M. Gohil and G. Joshi, "Perspective of Polycarbonate Composites and Blends Properties, Applications, and Future Development: A Review," in *Green Sustainable Process for Chemical and Environmental Engineering and Science*, Tariq Altalhi and Inamuddin (eds.,) Elsevier, 2022, pp. 393–424. doi: 10.1016/B978-0-323-99643-3.00012-7.

16. Thomas Xometry, *All about Polycarbonate Resins – Properties and Uses*, Thomas A Xometry Publication, 2022.

17. P. Raj and R. Kumar, "A Brief Review: Study on Mechanical Properties of Polycarbonate with Different Nanofiller Materials," in Rajmohan, T., Palanikumar, K., Paulo Davim, J. (eds.), *Advances in Materials and Manufacturing Engineering: Springer Proceedings in Materials*, Springer, 2021, pp. 285–291. doi: 10.1007/978-981-15-6267-9_34.

18. B. Demirel, A. Yaraş, and H. Elçiçek, "Crystallization Behavior of PET Materials," *Balıkesir Üniversitesi Fen Bilimleri Enstitü Dergisi*, vol. 13, pp. 26–35, Jan. 2011.

19. S. H. Park and S. H. Kim, "Poly (Ethylene Terephthalate) Recycling for High Value Added Textiles," *Fashion Text*, vol. 1, no. 1, p. 1, Jul. 2014, doi: 10.1186/s40691-014-0001-x.

20. M. Olam, "Pet: Production, Properties and Applications," in *Advances in Materials Science Research*, Wythers, M.C (ed.), Hauppauge, NY: Nova Science Publishers, 2021, pp. 131–162.

21. M. K. Akkapeddi, "Commercial Polymer Blends," in *Polymer Blends Handbook*, Dordrecht: Springer Netherlands, 2014, pp. 1733–1883. doi: 10.1007/978-94-007-6064-6_22.

22. K. Widiyati and H. Aoyama, "Coupling of Packaging Efficiency and Aesthetic Shapes for PET Bottles," in *Volume 2: 32nd Computers and Information in Engineering Conference, Parts A and B*, Utracki, L.A. (ed.), American Society of Mechanical Engineers, Aug. 2012, pp. 551–558. doi: 10.1115/DETC2012-70059.

23. H. Masoumi, S. M. Safavi, and Z. Khani, "Identification and Classification of Plastic Resins Using Near Infrared Reflectance Spectroscopy," *International Journal of Mechanical and Industrial Engineering*, vol. 6, pp. 213–220, Jan. 2012.

24. Y. Zhuang, S.-W. Wu, Y.-L. Wang, W.-X. Wu, and Y.-X. Chen, "Source Separation of Household Waste: A Case Study in China," *Waste Management*, vol. 28, no. 10, pp. 2022–2030, 2008, doi: 10.1016/j.wasman.2007.08.012.

25. M. T. Carvalho, C. Ferreira, A. Portela, and J. T. Santos, "Application of Fluidization to Separate Packaging Waste Plastics," *Waste Management*, vol. 29, no. 3, pp. 1138–1143, Mar. 2009, doi: 10.1016/j.wasman.2008.08.009.

26. F. Awaja and D. Pavel, "Recycling of PET," *European Polymer Journal*, vol. 41, no. 7, pp. 1453–1477, Jul. 2005, doi: 10.1016/j.eurpolymj.2005.02.005.
27. R. Navarro, S. Ferrándiz, J. López, and V. J. Seguí, "The Influence of Polyethylene in the Mechanical Recycling of Polyethylene Terephtalate," *Journal of Materials Processing Technology*, vol. 195, no. 1–3, pp. 110–116, Jan. 2008, doi: 10.1016/j.jmatprotec.2007.04.126.
28. K. Özkan, S. Ergin, Ş. Işık, and İ. Işıklı, "A New Classification Scheme of Plastic Wastes Based Upon Recycling Labels," *Waste Management*, vol. 35, pp. 29–35, Jan. 2015, doi: 10.1016/j.wasman.2014.09.030.
29. CANSA South Africa, *Plastic SA Plastic Identification Codes*, 2012, CANSA South Africa.
30. Wikipedia, "Injection Moulding," *Wikipedia*. https://research.tue.nl/en/publications/modelling-stretch-blow-moulding-of-polymer-containers-using-level
31. J. Huang, A. Veksha, W. P. Chan, A. Giannis, and G. Lisak, "Chemical Recycling of Plastic Waste for Sustainable Material Management: A Prospective Review on Catalysts and Processes," *Renewable and Sustainable Energy Reviews*, vol. 154, p. 111866, Feb. 2022, doi: 10.1016/j.rser.2021.111866.
32. X. Chang, Y. Fang, Y. Wang, F. Wang, L. Shang, and R. Zhong, "Microplastic Pollution in Soils, Plants, and Animals: A Review of Distributions, Effects and Potential Mechanisms," *Science of the Total Environment*, vol. 850, p. 157857, Dec. 2022, doi: 10.1016/j.scitotenv.2022.157857.
33. J. Gigault, A. ter Halle, M. Baudrimont, P-Y. Pascal, F. Gauffre, T-L. Phi, H El Hadri, B. Grassl, and S. Reynaud , "Current Opinion: What Is a Nanoplastic?," *Environmental Pollution*, vol. 235, pp. 1030–1034, Apr. 2018, doi: 10.1016/j.envpol.2018.01.024.
34. S. A. Mason, V. G. Welch, and J. Neratko, "Synthetic Polymer Contamination in Bottled Water," *Frontiers in Chemistry*, vol. 6, pp. 1–11, Sep. 2018, doi: 10.3389/fchem.2018.00407.
35. Don Rauf, "Microplastics in Bottled Water Are More Abundant Than Previously Thought," *Everyday Health*. https://www.everydayhealth.com/public-health/microplastics-in-bottled-water-more-abundant-than-previously-thought/#:~:text=Microplastics%20in%20Bottled%20Water%20Are, organs%2C%20and%20even%20the%20brain.
36. Aryn Baker, "Microplastics in Bottled Water at Least 10 Times Worse Than Once Thought," *TIME Magazine*, 2024.
37. J. Manzoor, M. Sharma, I. R. Sofi, and A. A. Dar, "Plastic Waste Environmental and Human Health Impacts," in *Environmental and Human Health Impacts of Plastic Pollution*, Harrison, R.M. (ed.), Pennsylvania, PA: IGI Global, 2020, pp. 29–37. doi: 10.4018/978-1-5225-9452-9.ch002.
38. J. M. Garcia and M. L. Robertson, "The Future of Plastics Recycling," *Science (1979)*, vol. 358, no. 6365, pp. 870–872, Nov. 2017, doi: 10.1126/science.aaq0324.
39. S. Dey, G. T. N. Veerendra, P. S. S. A. Babu, A. V. P. Manoj, and K. Nagarjuna, "Degradation of Plastics Waste and Its Effects on Biological Ecosystems: A Scientific Analysis and Comprehensive Review," *BioMedical Materials and Devices*, vol. 2, no. 1, pp. 70–112, Mar. 2024, doi: 10.1007/s44174-023-00085-w.
40. T. Saha, M. E. Hoque, and T. Mahbub, "Biopolymers for Sustainable Packaging in Food, Cosmetics, and Pharmaceuticals," in *Advanced Processing*,

Properties, and Applications of Starch and Other Bio-Based Polymers, Al-Oqla, F.M. and Sapuan, S.M. (eds.), Elsevier, 2020, pp. 197–214. doi: 10.1016/B978-0-12-819661-8.00013-5.

41. A. Ahuja, P. Samyn, and V. K. Rastogi, "Paper Bottles: Potential to Replace Conventional Packaging for Liquid Products," *Biomass Conversion and Biorefinery*, vol. 14, no.13, pp. 13779–13805, Dec. 2022, doi: 10.1007/s13399-022-03642-3.

42. L. Jiang, A. Gonzalez-Diaz, J. Ling-Chin, A. Malik, A. P. Roskilly, and A. J. Smallbone, "PEF Plastic Synthesized from Industrial Carbon Dioxide and Biowaste," *Nature Sustainability*, vol. 3, no. 9, pp. 761–767, Jun. 2020, doi: 10.1038/s41893-020-0549-y.

43. BIS, *BIS Standards*, Bureau of Indian Standards, 2024.

44. US FDA, *Determining the Regulatory Status of Components of a Food Contact Material*, US FDA, 2018.

45. EU, "Deal on New Rules for More Sustainable Packaging in the EU," *News European Parliament,* 2024. https://www.europarl.europa.eu/news/en/press-room/20240301IPR18595/deal-on-new-rules-for-more-sustainable-packaging-in-the-eu.

46. Dr Bharat Bhushan Nagar, "Is EPR an Effective Tool for Plastic Waste Management?," *360 Packaging*, 2020. https://www.linkedin.com/pulse/epr-effective-tool-plastic-waste-management-nagar-iosh-cwmp-iwm/

47. OECD, "New Aspects of EPR: Extending Producer Responsibility to Additional Product Groups and Challenges Throughout the Product Lifecycle", Working paper, 2023.

48. Rick Lingle, "Recycling Secrets of Sorting," *Plastics Today*. 2024, https://www.plasticstoday.com/packaging/recycling-secrets-of-sorting

49. K. Ragaert, L. Delva, and K. Van Geem, "Mechanical and Chemical Recycling of Solid Plastic Waste," *Waste Management*, vol. 69, pp. 24–58, Nov. 2017, doi: 10.1016/j.wasman.2017.07.044.

50. A. Samir, F. H. Ashour, A. A. A. Hakim, and M. Bassyouni, "Recent Advances in Biodegradable Polymers for Sustainable Applications," *Npj Materials Degradation*, vol. 6, no. 1, p. 68, Aug. 2022, doi: 10.1038/s41529-022-00277-7.

Index